Textbook of

TOWN PLANNING

A. Bandyopadhyay

Celebrating 64 Years 1960-2024
Since 1960 Books and Allied

Textbook of

TOWN PLANNING

Abir Bandyopadhyay

Prof. & Head, Department of Architecture,
National Institute of Technology, Raipur (C.G.)

BOOKS & ALLIED (P) LTD.

No. 1E(1), 'Shubham Plaza' (1st Floor)
83/1, Dr. S.C. Banerjee Road, Beliaghata, Kolkata 700010
Tel : (033)23740062, *e-mail* : booksandallied1960@gmail.com
website : www.booksandallied.com

Textbook of TOWN PLANNING

First Published : January, 2000

Reprinted : 2010
Reprinted : 2017
Reprinted : 2022
Reprinted : 2024

Publisher : Ruba Sen
BOOKS & ALLIED (P) LTD.
No. 1E(1), 'SHUBHAM PLAZA' (1st Floor)
83/1, Dr. S.C. Banerjee Road, Beliaghata, Kolkata 700010

Printer : EUREKA
91/A, Baithakkhana Road, Kolkata 700009

ISBN 978-81-87134-65-7

Price : **₹ 400.00**

Dedicated to :

my Parents

Smt. Nomita Bandyopadhyay.
Shri. Ansuman Bandyopadhyay.

PREFACE

Developments of cities and towns are as old as the civilization of mankind. Towns started developing as centers of commerce and as units of defense against alien forces. Slowly and gradually, in due course of time, towns started developing as a center of business and industrial activities, and also as a hub of administration. Thus economics started playing the key role in the development of the cities. Lately, we also have started understanding that all of our 'developmental' activities play a role in shaping and reshaping our environment, which are sometimes very detrimental for our future generations. Today, Planning has become a science of 'allocating chunks of scarce resources in a pre-conceived manner'. Resources, like natural wealth, money, time and work force are available in limited quantities and it is upto the Planners to decide how to use these resources appropriately for the betterment of our future, by exerting the least amount of stress on the environment. Though planning alone will not solve the problems of the Urban community, neither it can control the dilapidation of the scarce natural resources, but a study of planning will definitely help to generate a consciousness of the ills of irrational usage of the natural resources.

This book is my part of contribution to the students studying the subject of Town Planning in various universities in India as the future architects, engineers and the town planners will play a very important role in shaping the fate of our country.

Although written to serve as a text and a reference book for the students of town planning, the book will also be of special value to professionals. It will serve to systematize instructions in the planning field given through departments of Architecture, Civil Engineering and Urban Planning and is compiled to match with the syllabi of most of the Indian universities. The current universality of the planning profession has broadened the horizons for the persons in or entering the field. Professionals, now, cannot only restrict themselves with the regulations of landuse or building bye-laws, but also they must understand the implication of all the regulations of a proper development system.

To give them a better understanding of this system, they must know the history — how the ancient cities were made and how the cities grew in the past. Learning the positive and the negative aspects of the development of the towns in the past, they will better understand how and why certain things are practiced at present. Thus, they can apprehend the business of Town Planning in a better way and consequently they will be capable to take the decisions of developing our future cities in a proper manner.

This book will serve the purpose of all who are interested to know the business of city planning. Matters of this book are written point-wise, in a concise manner, so that it can serve as ready reference for anybody. Instead of going to greater depth of any topic, the book features important aspects of town planning to give a general insight of the subject to the reader.

For sake of convenience, the book has been presented in three units; the first deals with the History of Town Planning; the second with the Broad Processes of Town Planning as practiced in India and abroad, and the third deals with Rural Planning.

The important theories of town planning, which are worth mentioning, are included in second unit to give an idea of the present (and the past) trends in the thoughts behind the development of the cities.

Policy making, importance of management systems, importance of the legislations related to Planning are also added to give a better insight of the planning system. Town planning is in actuality a combination of various other streams of specializations like demography, social sciences, economics, management systems etc. But in this book stress is mostly given to the 'Physical' part of planning which includes Transport Planning, Housing, Neighborhood Planning and the matters related to the Inner Cities of the Metropolis.

A chapter on Rural Planning has been added at the end of the book to establish the fact that countries like India cannot develop if the spine of our civilization, i.e. the villages, are neglected in the developmental process. Planning of the villages in India are extremely important to stop the in-migration of the people to the bigger towns, and the chapter highlights the various actions taken up by the Government of India to improve the condition of our rural areas.

I express my indebtedness to the authors of the books on this subject to which references have been made. I am extremely grateful to all my teachers, friends and men in my profession who have inspired me to produce this book.

I am grateful to my colleague Ar. Sameer Ghorpade, my students Sh. Jaiprakash Keshwani, Sh. Prithviraj Sengupta and Ku. Deepanwita Chakravarty for their contribution for making the sketches and drawings.

Last, but not the least, I am grateful to my wife, Preeti, who kept on giving me important suggestions during the writing of this book.

01.12.1997
Qtr. E-16
Govt. Engg. College Campus
Raipur - 492 010 (M.P.)

Abir Bandyopadhyay

CONTENTS

Unit-I

Unit-II

Unit-III

Unit-IV

INDEX OF DRAWINGS

If we could know where we are and wither we are tending, we could better judge what to do, and how to do it.

Abraham Lincoln.

Unit I

History and Development of Settlement Planning

PRINCIPAL ASPECTS FOR CONSIDERATION IN THE DEVELOPEMENT OF SETTLEMENT PLANNING THOUGHT

Ever since man has felt the necessity to stay in a group, small cluster of houses started coming up which slowly and gradually took the shape of small villages. Agglomeration of many such villages and increment in the sizes of the villages lead to the development of towns. There are several factors which are responsible for the location, development and the evolution of towns in the history of mankind. Several towns have been made to serve the various purposes of human needs and a study of the development of the towns reveal to us the factors which have played an important role in the growth of our civilizations. These factors, even today, play an equally important role in the shaping up of the present day towns.These factors may be listed as :

1. GEOGRAPHICAL FACTORS which include:
   - **(a)** Earthquake and floods.
   - **(b)** Land characteristics, fertility etc.
   - **(c)** Water availability, climate, weather etc.
   - **(d)** Resources.
2. ECONOMIC ASPECTS which include:
   - (a) The Production system like
     - (i) Agriculture.
     - (ii) Industry.
     - (iii) Trade and commerce.
   - (b) The economic leadership.
3. POLITICAL ASPECTS which include :
   - **(a)** Defense considerations.
   - (b) Administration.

4. SOCIO-CULTURAL ASPECTS which include :

   (a) Castes, class distinction.

   (b) The poor and captives.

   (c) Mysticism.

   (d) Cultural pattern.

   (e) Social Grouping.

5. TECHNOLOGICAL PROGRESS which includes :

   (a) Influence in industrial production.

   (b) Building and construction works.

   (c) Transportation within the settlement and beyond.

   (d) Utilities and services.

The history of human settlements have been considered by the planners predominantly based on the development of civic design, but now, town planners have started studying the history of human settlements not only from civic design point of view but also from the above mentioned points of view, as the general process of evolution of the Human race is reflected in the development of the towns.

1 EARLY CIVILIZATIONS

1.1 INTRODUCTION :

It has been observed that in the early days most of the civilizations have sprung up on the banks of some navigable river or at any natural port which not only provided security to the town but also acted as a communicating link from one civilization to another.
Thus, physical safety and communication links are the two prime factors which gradually lead to the development of the towns.

Politically, the history of civilization were dominated by mainly two types of people namely the King or the Rulers and the Priests. The rulers had the support of the armed forces and the priests had the sentimental support of the people at large. Hence, we see, that when the kings and the priests performed their duties harmoniously, civilizations flourished and when there were any discrepancy amongst them, conflicts and confusion started which lead to the ultimate downfall of that civilization.

The earliest civilizations were seen on the banks of Euphrates and Tigris, Nile, Ganga, Sindhu (Indus), Yamuna, Yang-ste-kiang etc.

1.2 SUMERIAN CITIES:

Sumerian cities flourished from 12000 B.C. to 3000 B.C. The Sumerian City was protective and compact in form. The buildings were developed horizontally which adapted well to the climate. The market was situated near the center and the residences were placed all around.

The development was symmetrical taking the river as the central axis.

Transportation was also done by water .The houses were predominantly made of timber. The skyline of the cities were dominated by Ziggurats which were sometimes 8 stories high, (vide Fig. No:: 1 UR-Babylon)

1.3 EGYPTIAN CITIES:

The Egyptian civilization grew only on one side of the river as the river Nile was very wide for navigation. The peoples' life was dedicated to the Pharaoh, whose artistic ambitions and orders build up the towns. The towns were mainly for the slaves and the artisans for making the

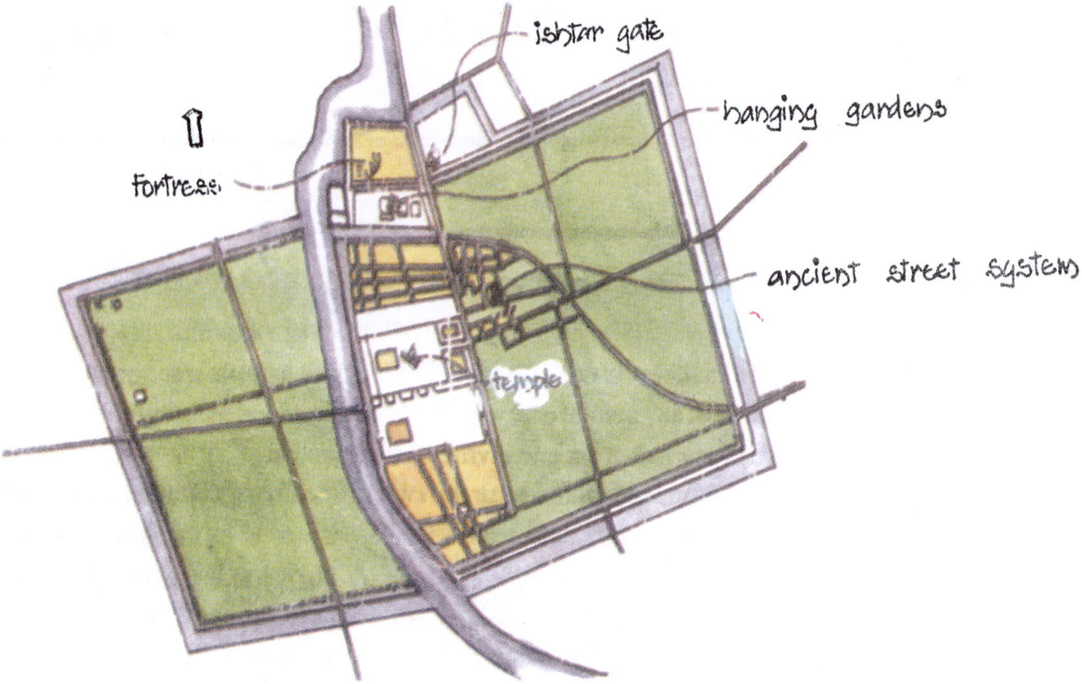

Fig. No. 1(a) : *City of Babylon*

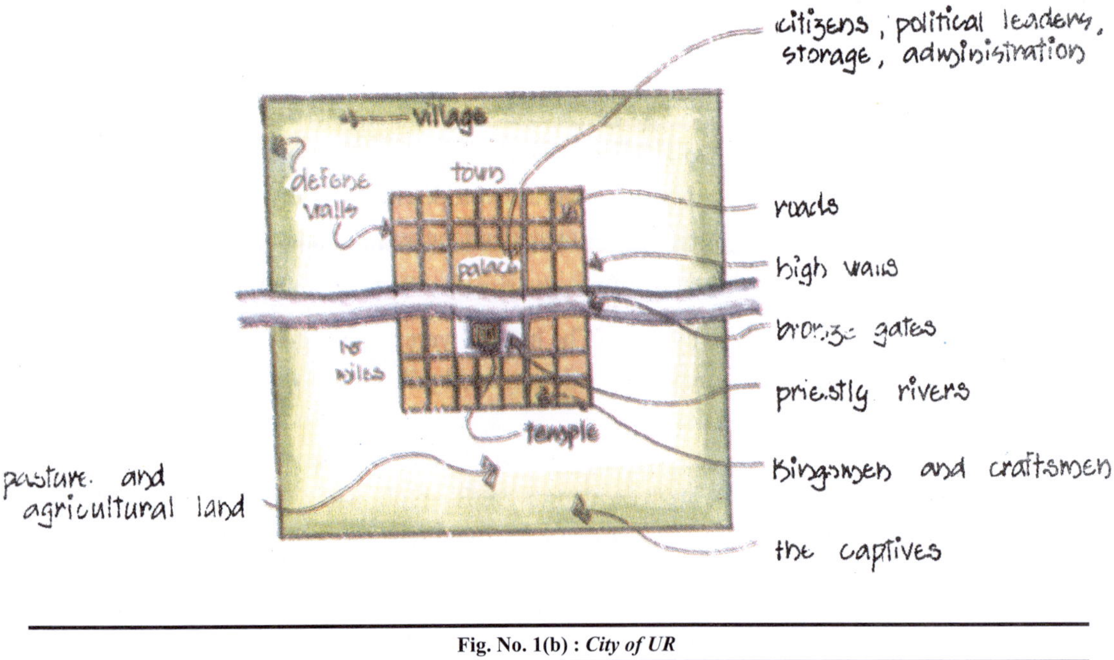

Fig. No. 1(b) : *City of UR*

Pyramids. The 'cells'(or rooms of the houses) were arranged around a common courtyard and the buildings were made of sun dried bricks. The narrow lanes served as open drains and walkways. Walls surrounded the cities for protection against the floods of the river Nile.

The dwelling were of different types and sizes.The most important commodities were placed on the upper right corner of the towns.

The roofs of the houses were made of reeds covered with mud plaster.

A ventilating device known as 'mulguf was used on the roofs. In this system warm air is passed through a wind catcher placed at the roof to the basement which was filled with water. The air passing through this water got humidified and then this was passed through the perforated flooring of the ground floor to make the rooms cool.The wind catcher was made rotating to catch the changing direction of the wind. This system is still seen in some parts of Iran where it is known as 'Badgir'.

In the second millennium B.C., the Egyptians made great temple cities on the bank of the river Nile. These cities had monumental avenues, colossal temple plazas, rock-cut tombs etc. and are seen in Memphis, Thebes, Tel-el Amarna. In Thebes there is the dramatic avenue of sphinxes and at Tel-el Amarna the temple complex was 1/3 miles by 1/2 miles.

vide Fig. No.: 2 — City of El-Kahun.

1.4 GREEK CITIES :

During the Greek period the 'City States' were flourishing. By 7th.Cent. B.C., the nobles became the rulers and Athens came under the rule of tyrants. Consequently land holding peasant classes developed. By 5th. Cent. B.C., democracy took root in Athens. Magistrates were elected, philosophy was nurtured, science was investigated—an environment of culture was developed by people like Socrates, Plato and Aristotle.

The Greek towns were divided into three main physical parts: one for the Gods, one for administration and the other for the dwelling houses. The city plan was conceived to serve all the three classes of people namely, the craftsmen, workers and soldiers. The towns were designed to satisfy the requirements of hygiene, defense and circulation. The temples were built upon Acropolis. Most of the Greek cities had the following characteristics:

(i) Wandering and unpaved lanes.

(ii) No drainage and sanitation system.

(iii) Water was got from the local wells.

(iv) The wastes were disposed off on the streets.

(v) There were no palaces, even the public buildings were less apart from temples and a common assembly place known as PNYX and an irregularly shaped market place or AGORA.

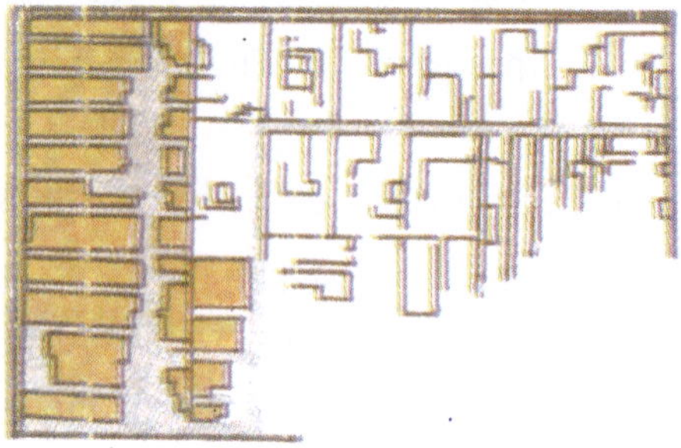

Fig. No. 2(a) : *City of El-Kahun*

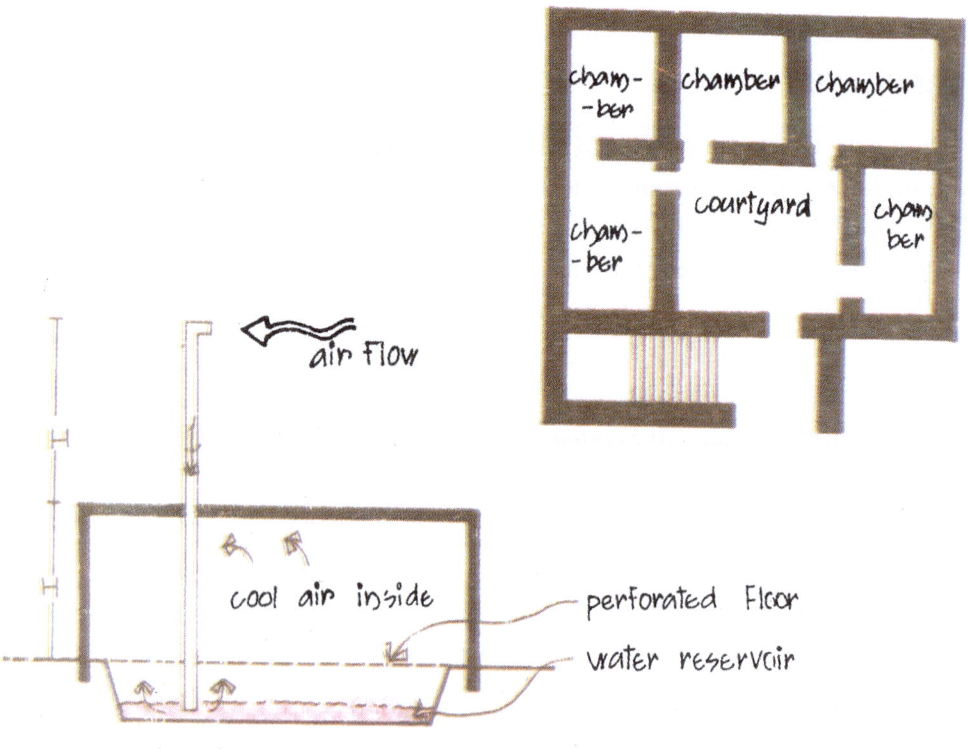

Fig. No. 2(b) : *Mulguf System*

(vi) There were no class distinction in the society.

(v) In the houses, the rooms surrounded the courtyards having windowless facades, vide Fig. No. 3 — City of Athens.

In the later part of 5th. Cent B.C., Hippodamus, an architect from Miletus, advocated an outstanding theory in the history of town planning by applying the GRID IRON pattern of road layout. This system was introduced to rebuilt the Greek cities which were destructed by the Persian invasion. For laying out the perpendicular roads, individual buildings were made the common denominator and the functions of the buildings were given importance. The rough topography of the Greek cities lead to creation of steps on the roads, which did not create much problem as most of the movement were on foot.

Public spaces played a very important role in the life of the Greeks.The various types of public spaces were :

AGORA or the Market place, was situated at the center of the city .The space occupied by agora was approximately 5% of the city area and the dimensions were nearly l/5th.of the width and breadth of the whole town. The Agora was generally faced by the assembly hall, council hall, and council chamber (PRYTANEUM).

The Greek cities made in the later part of the 5th.C. B.C.,had some sort of building bye-laws, by which the projection on the first floor were restricted and sanitation was given importance. The population size of the cities were guided by the amount of food and water supply. The population of Athens was about 40,000 and that of other cities ranged from 10,000 to 5000. The climate determined the orientation of the buildings. The rooms faced a southern private courtyard where a projected colonnade protected the rooms from direct sunlight. The north wall was punctured with small windows.

vide Fig. No.4 — City of Miletus.

During the Hellenistic period, Greece came under the rule of the Macedoneans. Many public buildings were made during that period namely, ODEION, TREASURY, LIBRARY, PRISON etc. were added to AGORA, Baths.Stadia etc.

1.5 ROMAN CITIES :

Roman civilization flourished from the 6th.C. B.C. to 5th.C. A.D. The Romans lacked creativity of the Greeks but built in a grand scale. They developed water supply system, drainage system etc. Water was brought from far away places by the means of aqueducts. Public buildings like baths were more developed during this period.
The Romans founded their towns mainly on three occasions :-

(a) When the towns became colonial çenter.

(b) When it was made a local capital.

(c) When it became an imperial residence.

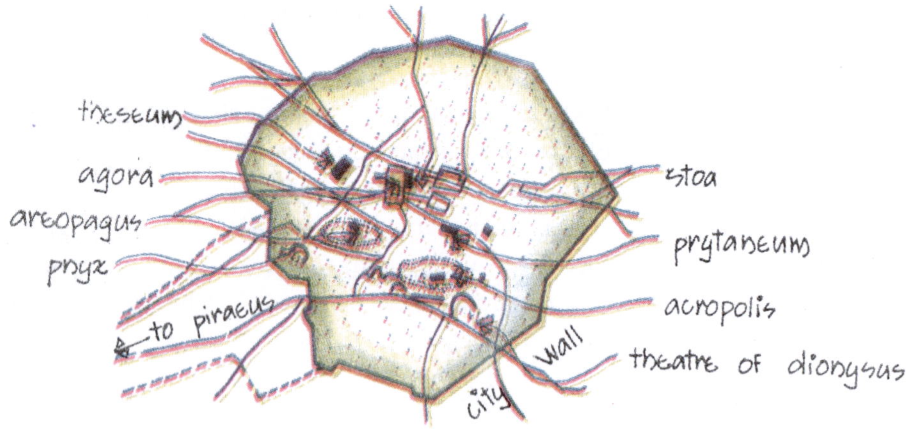

Fig. No. 3 : *City of Athens*

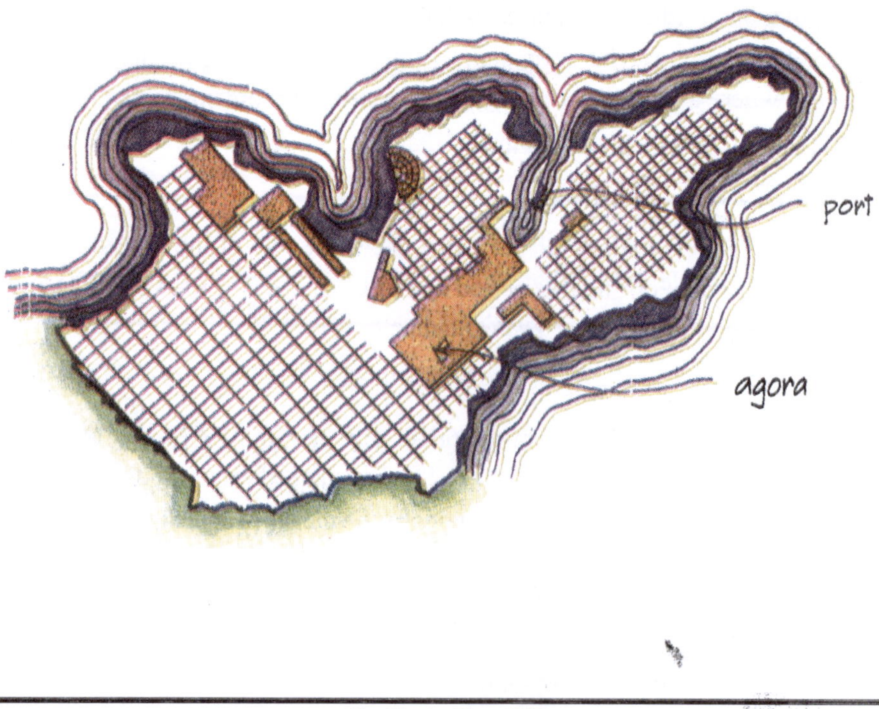

Fig. No. 4 : *City of Miletus*

The towns were laid out to comply with fulfilling the religious auspices, defense requirements, and technical considerations.

(A) The religious auspices include:

- To establish a city wall first. There were different religious activities within and outside the city wall.
- Next of importance were the two cross roads; DECUMANUS — running east-west, and CARDO running north-south. The main road running north-south is known as CARDO MAXIMUS. This system of road layout is known as CENTURATIO SYSTEM. FORUM was to be developed at the crossing of the main roads.
- The rest of the town was to be divided into square or rectangular plots of one side equaling 120'.(36.57 m).

(B) The defense requirements include :

- A defense wall.
- Straight roads to avoid curvatures in streets which enabled to perform military parades easily.
- Long distance visibility for easy policing required the roads to be absolutely straight.
- The roads were at least 8' (2.40 m) wide.

(C) The technical considerations include:-

- Setbacks of 2'6" (0.76 m) from the boundaries of each buildings.
- Drainage and water supply were elaborately prescribed.
- All construction within the towns was to be in concrete and stone. In specific cases they were faced with bricks.

The Roman towns were mainly divided into 3 categories.

(i) COLLONIA : Which were high ranking cities having all benefits, where there as hierarchy of plots for people having various social status.

(ii) OPPIDIA : Towns which were conquered from others.

(iii) CASTRUM : Military camp towns,

vide Fig. No. 5 —Roman Towns.

1.6 MEDIEVAL CITIES:

Five centuries after the death of Christ, the Roman Empire crumbled under the pressure of luxury, pomp and ceremony. The Roman capital was shifted to Byzantium (Constantinople). The urban population returned to rural life and there was an era of social and economic confusion. By 7th.C. A.D., there was a rise of Muslim culture under prophet Mohammed. Muslim religion started spreading all over Europe and Africa and by 10th C. A.D.,it was challenging Christianity. Separate types of Muslim towns were coming up, characterized by gardens, fountains and topiary. After the fall of Roman Empire, independent city states developed with average population between 4000 and 12,000. Rivalry between the cities lead to fortification of them. Insecurity of life lead to the importance of religion. Monasteries served as refuge of the oppressed and hence the church strengthened its position. Insecurity of the unprotected places gave the impetus to the people to come back to the urban life. This brought revival of trade, and the need of self sufficiency within smaller city states lead to diversification of the industrial production. To permit whatever inter-settlement trade still possible, commercial areas developed near the gates of the fortified walls so that the outsiders would not go into the interior areas. A feudal system of government developed and a new form of freedom emerged.

Guilds were formed of the merchants and craftsmen like those of weavers, tailors, masons, metal works, carpenters, leather works glass makers etc., who established regulations to control their production, maintain their prices and protect their trade and were having a new social order in the making.

A wealthy mercantile class were rising to challenge the power of the feudal lords. The city council or municipality emerged. Social stratification and grouping developed as :

The nobles and rich merchants near the entrance gates of the defense wall of the city, and the craftsmen and agricultural serfs in the rest of the walled city but with no distinct zoning among them.

The religious leadership was at the highest level of land near the center of the settlement.

The City Council, as a symbol of the people, was near the geometrical center of the settlement.

The early towns were dominated by the church, or monastery, or the castle of the feudal lords. Church plaza, market plaza, town hall and guild halls were made adjacent to it.

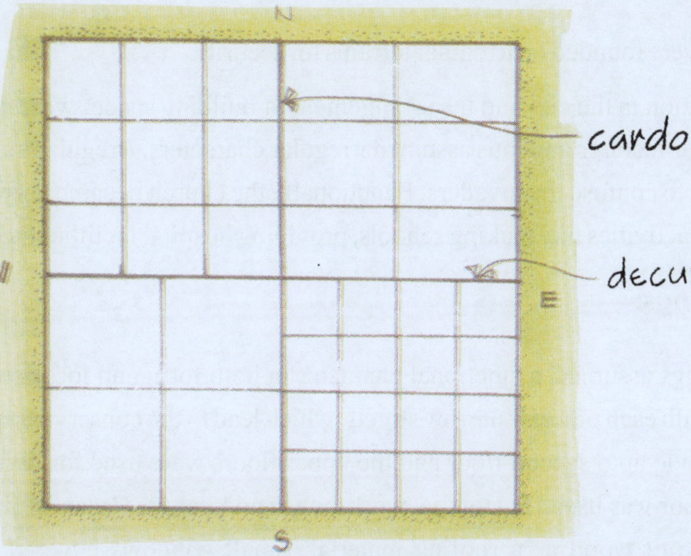

Centuratio system

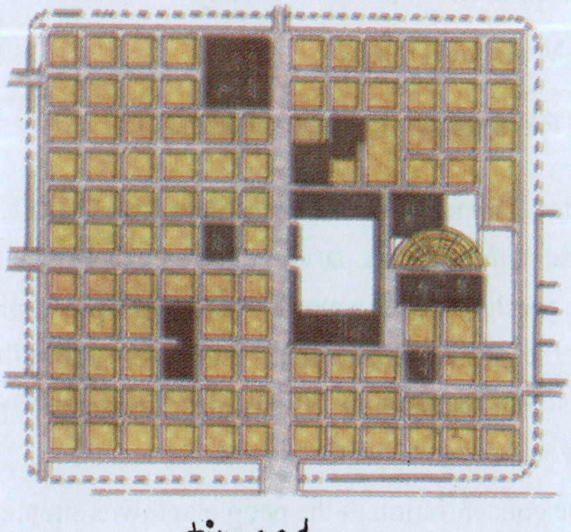

Fig. No. 5 : ***Roman towns***

1.6.1 The Cities:

The cities were founded on irregular terrains for security.

The circulation in the city and the arrangement of building spaces were molded to match with those irregular features and thus assumed irregular characters. Irregular road patterns were consciously made to confuse the invaders. Functionally the church became very powerful and it undertook certain activities like making schools, providing hospital facilities etc.

1.6.2 The Buildings:

The buildings assumed a functional character in both form and location. The buildings were connected with each other by narrow streets, which lead to the conservation of privacy. The buildings were made upto second floor and the upper floors were used for living and sleeping and the ground floor was used for storage, workshops and kitchen. Cases of fire in some cities prohibited the use of wood as a roofing material. Small windows covered with glass and parchment papers were used.

1.6.3 The Institutions:

Monasteries and guilds joined hands together for creating trained personnel in the fields of law, medicine and arts. This lead to the birth of universities.

Art was introduced in city planning, like the paving of squares and streets. A 'Human Scale' pervaded in the informal environment of the city.

vide Fig. No. 6 — Medieval Cities.

1.7 NEO-CLASSIC OR RENAISSANCE PERIOD :

Renaissance started in mid 14th C. and lasted upto 17th.C. During this period the medieval cities disintegrated into smaller parts and reduced the contact between one country and another. The towns had a few hundred to a maximum of 500,000 people. The physical size of the towns were restricted by the girth of the fortification. Towns were built within the ready reach of one another—in a day' s walking distance, i.e. 20-30 miles. There was no system of sewage disposal and the streets were used to dispose off the wastes.

Trades brought the concentration of the people to towns situated of the main cross roads. Ownership of the lands shifted to the merchants and the power of the feudal lords diminished.

Printing press was invented and ways were devised to improve the simple hand machines. Gun powder was invented in the 15 th. C., and new techniques of warfare were introduced, which

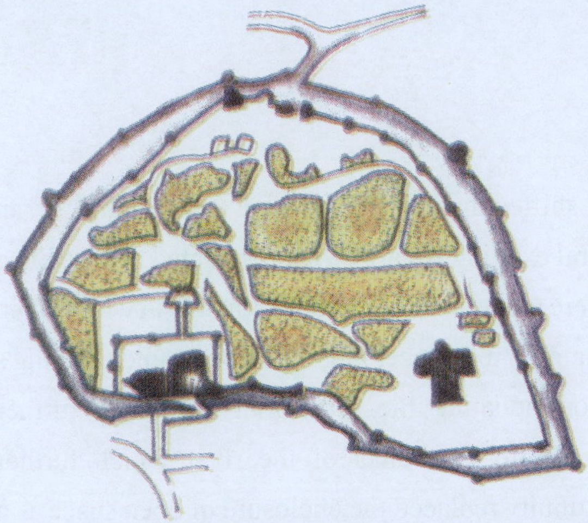

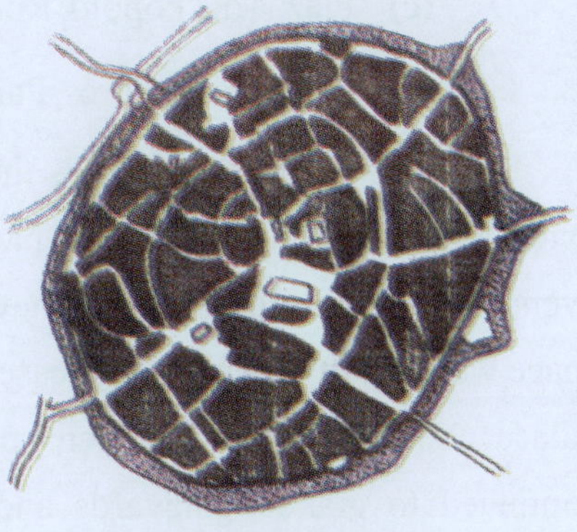

Fig. No. 6 : ***Medieval Cities***

changed the strategies of warfare and old fortifications were found inadequate. Thus military engineering became a science.

The contrast between the rich merchants and the poor increased and hence, the insecurity of life of the poor also increased. As a result religion again became very important. Church was controlling all the fields of knowledge which helped in the revival of classical knowledge. It became the ambition of the rulers to display their ambitions and power. This display and exhibitionism were manifested in the construction of formal and monumental buildings, drawing upon the classical heritage of Rome. Mainly two types of manifestations are noticed in this period :

(i) New towns were found in which the central and the most dominating buildings were those of the nobles, i.e. the courts of the kings. Here religion was subservient to the exalted position and the glory of the nobles. For example, Versailles in France, Canberra in Australia and Washington D.C. in U.S. A. major parts of London in U.K. was designed by Christopher Wren.

(ii) In the old medieval towns that existed, development was done in the designing of the public congregation places like the squares, piazzas (plazas) — in and around religious buildings and developing wide avenues curving out of wide cramp areas and leveling spaces of old defense walls; e.g. —

(a) The piazza of St. Marks, Venice.

(b) Piazza of St. Peters'.

(c) Piazza del Popolo, Rome.

(d) Place des Victores, Paris.

In the renaissance period, the design shifted from the enclosed architectural to an extension and expansion of open spaces. Several existing squares were connected by tree lined avenues as in Paris. In some cases new plans were made and executed for squares in which open space was almost completely released. For example in France, large garden around the king's palace continued to large open spaces accommodating the secretariat buildings; further continued to grand boulevards and still further to the banks of the river which further exemplified the continuity of space. Thus, continuity replaced the enclosure of open space as a new direction of civic design.

vide Fig. No. 7 — St. Marks piazza and other renaissance city squares.

1.8 BAROQUE CITY:

The axial system of Planning which was introduced by Lorenzo Bernini during the Renaissance period was developed during this time. King Louis XTV ordered to remove his palace from the congested Paris to the open hunting ground of Versailles and ordered to have the avenues to radiate out from this magnificent palace.

After Napoleon III rose to power in 1853, the-cities were congested with slums and the condition of Paris was deteriorating. Mechanical traffic was to be introduced on the roads and it was an urgent necessity to check the haphazard growth of Paris. George Eugene Hausmann came up with the novel plan of having straight avenues joining the important places; boulevards were made and some form of building bye-laws like height restrictions were introduced.

The main feature of Baroque planning were as follows:-

(a) Avenues, (b) Fountains, (c) Axis and (d) Geometry.

Example: The SHONE BRUNN Palace at Germany where the sides of the trees were also chopped off along the road to achieve the 'axis' of the design.

Thus we see that in the Medieval Period, The main emphasis were given to the 'Mass' of the buildings, in the Renaissance period the importance was given to the 'Space' and in the Baroque period, the importance was laid upon both 'Mass' and 'Space'.

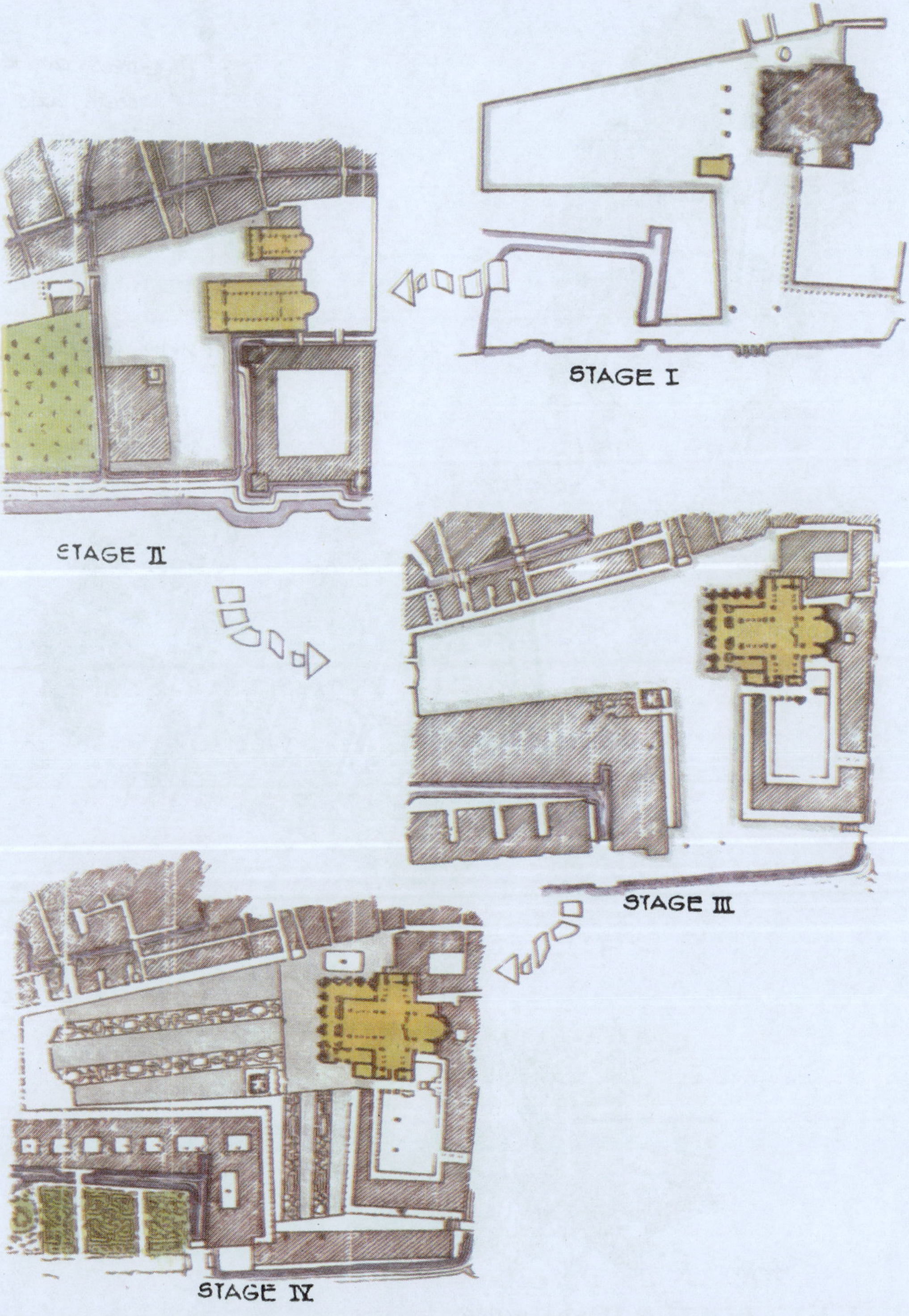

Fig. No. 7(a) : *St. Marks Piazza & its development*

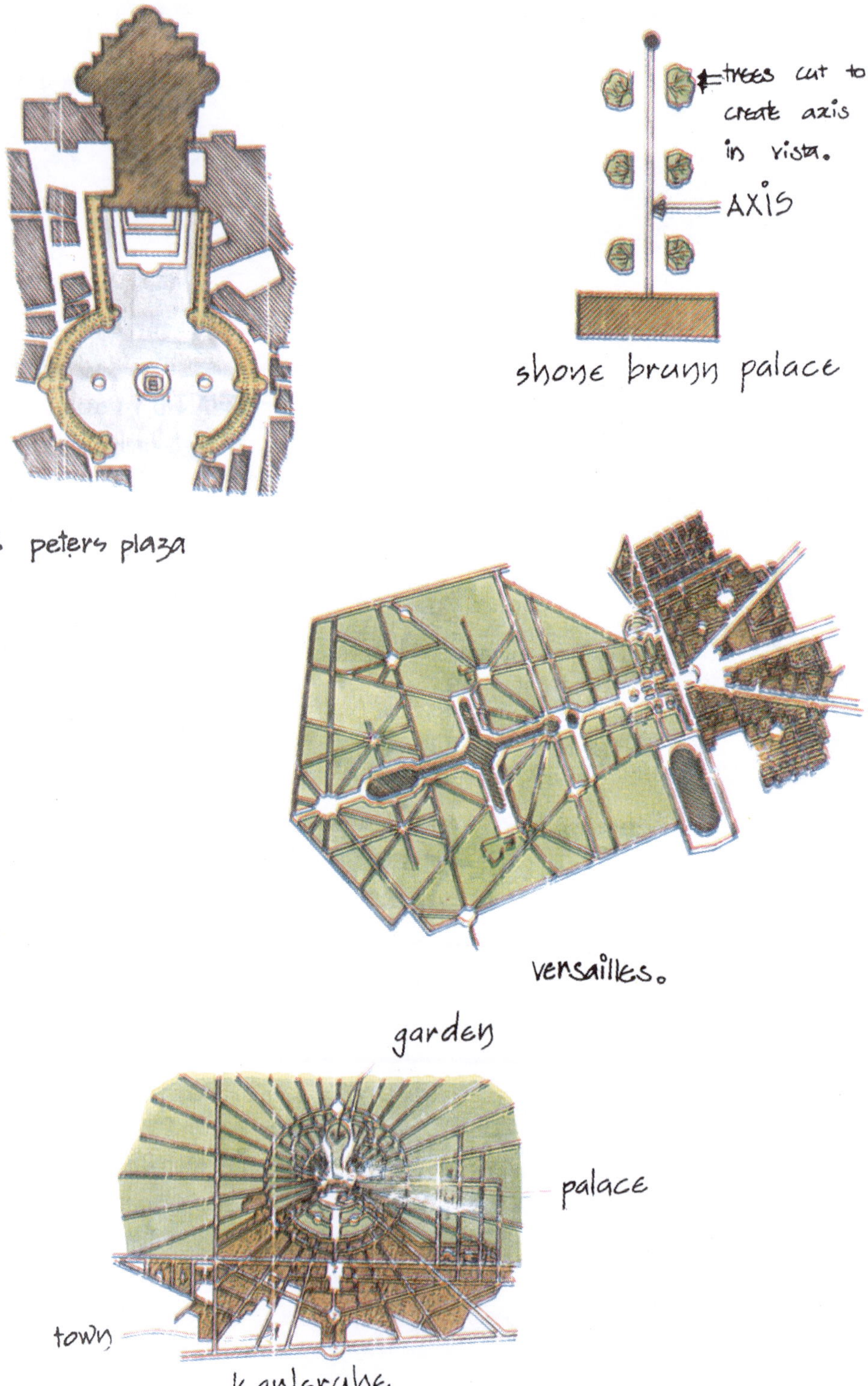

Fig. No. 7(b) : ***Other Renaissance City Square***

2 HISTORY OF TOWNS IN INDIA

2.1 INTRODUCTION:

The Hindu culture which developed in India was centered around the four Vedas namely, Rik, Sama, Yajur and Atharva. The Vedas were also divided into sub-parts known as 'Vedangas', 'Vedantas','Upanishads' and 'Puranas'.

According to the 'Vedas' there are four ages ('yugas') of human existence on earth namely, Satyayuga, Tretayuga, Dwaparyuga and Kaliyuga. The Vedic Period started anytime between 15000 B.C. and 3500 B.C. The Ramayana war took place around 12000 B.C. and the Mahabharata War took place around 3000 B .C. (According to Aryabhatta it took place in 3102 B.C.).

During Rigveda the rivers Indus and Saraswati were largely mentioned, Yamuna was mentioned occasionally, but no name of Ganga were there. Scientists say that there might be a marshy land in the gangetic plane and to fill up the land it must have taken 15000 to 20000 years; hence the Ramayana period was roughly 15000 years old after the starting of the Vedic Period.

At the starting of the Vedic period, there was no need for security as the Indus plane was fully secured. Hence no technology developed. No proof of any permanent construction is got and the people lived in small communities as per Manasara. There were no knowledge of geometry as angles were not known and space was measured by Purusha — an equilateral triangle of 1'6" (0.365 m) side. All the settlements were located on good alluvial soil and near a water body of a tank, lake, canal, river etc.

2.2 THE INDUS VALLEY CIVILIZATION :

Carbon tests of the remains that were got from the Indus Valley reveal that the civilization flourished 3000 years B.C. Some say that the cities like Mohenjodaro, Harappa, Nasik, Tamluk etc. were devastated by the Mahabharata war. Historians claim that there was an identifiably independent culture at Mohenjodaro and Harappa that were marked by development of wheels, use of metals, knowledge of arithmetics, use of coinage, art of expression and representation, architecture and planning techniques. Evidence of the existance of any other similar civilization belonging to the same era is yet to be found. Knowledge of geometry, angle, rectangle, slope of ramp, circle, helical coil etc. were also known to the people. Circular tubes for drainage and water supply were also in use.

2.2.1 Mohenjodaro:

There were no fortification and the major streets were laid in north-south direction. Other roads cut these roads at right angles and the streets within the built-up areas were narrow. Zoning was distinct for distinct groups. The city was divided into three distinct groups or divisions:

(a) The religious, institutional and cultural areas around monastery and great bath in the western part including the temple.

(b) The northern part was principally for the production of agriculture and industries.

(c) The southern part principally for the administration, trade and commerce at the meeting of the east road and the first street near the palace.

Salient Features:

(1) The construction technique was very well advanced, buildings were made of sundried bricks.

(2) Buildings ranged from two rooms to mansions of many rooms.

(3) Underground drainage and sewerage systems were also provided to the houses.

(4) Helical pumps were used to pump water in the great bath.

(5) The principal buildings were the monastery and bath, indicating the influence of religion as the main source of the development of the culture and defense was felt not as important.

(6) There was an all round development of arts and crafts, science and technology. Thus the ruins show an excellent way of adjusting to the contours.

All these show that there was a very advanced civilization,

(vide Fig. No. 8 — The City of Mohenjodaro).

2.3 THE BUDDHIST PERIOD:

This period corresponds to the Greek period of Europe. Faith in hindu philosophy was declining and a new philosophy was coming up which laid stress in the making of Buddhist prayer halls and stupas. In the fourth Cent. B.C. Alexander the Great invaded India and there after a commercial link between the Greeks and the Indians developed. A great school of art known as the Gandhara School developed which was Greek in flavor and Indian in content. There after two memorable reigns followed which have contributed to the glory of Indian culture. The first one of these were MAURYAN PERIOD from 324 B.C. to 187 B.C. This was followed by an era of political instability of 500 years after which the golden age of GUPTAS came into rule from 320 A.D. to 520 A.D.

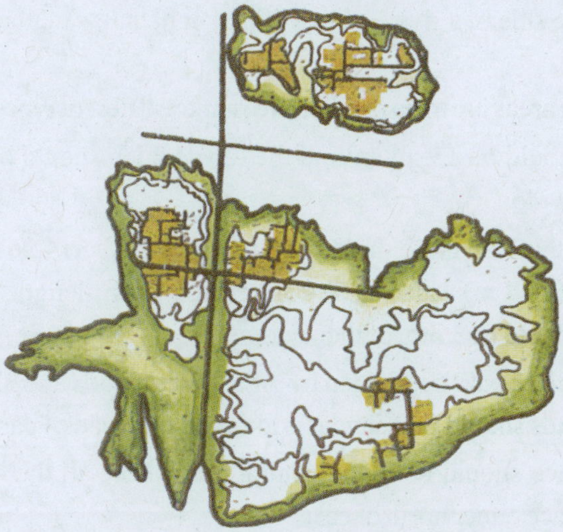

Fig. No. 8 : *City of Mohenjodaro*

2.4 THE MAURYAN PERIOD :

King Chandragupta was the founder of this dynasty who had Kautilya (or Chanakya) as his Prime Minister. Chanakya was the author of ARTHASHASTRA —containing discourses on many subjects, inter-alia duties of government, administration, town and country planning and housing.Some interesting extracts from Arthashastra is given hereunder :

- Towns should not have surplus population — extra population should be shifted to new areas to make new settlements.
- The towns should be located in such a way that they can help one another at the time of danger.
- There was also a system of tax relaxation for a few years for the people who come to stay in a new town.
- Markets were proposed to be placed at highways and road crossings to improve commerce.
- The new towns should have a higher proportions of agriculturists and Shudras.
- Dams should be constructed, temples should be provided and arrangements should be made to look after the aged, children and the infirm persons.

2.4.1 Town Planning :

- A city has to be located at the central portion of the country to facilitate trade and commerce.

- There should be a continuous source of water supply and the site of a town should be by the side of a river or an artificial or natual water body which do not dry up easily.
- Separate areas are to be provided to market different types of goods.
- There should be a wall around the town which should be 6 dandas high and 12 dandas wide.
- Beyond this wall there should be 3 moats of 14 feet (4.26 m), 12 feet (3.65 m) and 10 feet (3.04 m) width to be constructed 4 arm-lengths apart.The depth of the moats should be 3/4th. of the width.
- The towns should be divided by 3 East-West roads and 3North-South roads. The main roads should be 8 dandas wide and other roads 4 dandas wide.
- The palace should be in the central part where all the 4 varnas were living. It should face either north or east.
- The houses of the priests and the ministers should be on the south-east; traders, skilled workers, and Kshatriyas on the east; the treasury, goldsmiths and the industries on the south; forest produce, armaments on the north-east; and doctors, city fathers,rmy commanders, artists on the south. The oppressed classes should be housed beyond the cemetery.
- There should be one well for every group of 10 houses.

2.4.2 Housing :

- Foundation of every houses should be two arms-length deep.
- Every houses should be provided with latrine.
- The open spaces in-between two houses and two balconies should be 4 foot-length wide (1.2 m).
- There should be no obstructions to the door shutters opening inside.
- To ensure lighting in the rooms a small window should be placed high up in the wall.
- The roofs should stand rains.
- Bathrooms, latrines, grinding place and open spaces excluding special halls and rooms must be made available for the common use of all tenants in the house.
- Those who will obstruct steps, nahanis, or cause any damage to the walls of buildings due to sullage water and dirty water will be punished.
- A drain must be provided to carry out the waste water, it should have an adequate velocity.
- Those who throw garbage on streets should be punished.

- A census should be taken of all men and women and information about the caste, gotra, occupation, income and expenditure should be noted down.
- For summer seasons, arrangements should be kept ready for attending to extinguishing of fire.
- For festivals special bathrooms and toilets should be constructed, water should be drained by means of a nullah.
- The nahani should be 3 arm-lengths wide.
- Separate spaces of 3 foot-lengths wide and four foot-lengths long should be provided or locatig water for fire fighting, grinding and pounding arrangements.

Thus it bcomes clear that in his book 'Arthashastra', Chanakya consciously gave stress to not only the physical aspects of the towns but also the social and economical aspects which act as the most important things for the survival of a town.

2.4.3 Samrat Ashoka :

Ashoka reigned from 273 B.C. to 236 B.C. During this time the Mauryan Empire reached its maximum glory. He turned into a Buddhist and made many Buddhist Viharas. He also engraved the principals of his religion on pillars of rock and in caves. He also dug wells, planted trees, built rest houses and hospitals.

The Mauryan Empire was divided into provinces, cities and municipalities. The village administration was in the hands of 'gramicas'. For 10 villages there was a 'Gopa' and above it a 'Sthanaka'. Taxes collected from the citizens were spent on public welfare works like tree plantation, digging of lakes, and constructing rest houses and hospitals.

2.5 THE GUPTA PERIOD :

During the Gupta period the Buddhist religion fell down and again Hinduism was revived. The reign of Samudragupta and Chandragupta II, also known as Vikramaditya, is looked upon as the Golden Age of India. This period corresponds to the Roman period of the west.

2.5.1 King Vikramaditya :

Vikramaditya was a great patron of learning and Sanskrit reached its glory under the poets like Kalidasa during his period. Cities like Ujjain, Pataliputra, Mathura and Sharavati became great centers of learning and culture. Fa Hien visited India during this time and he writes that the economy was sound and the government was liberal. The people had enough foods, clothes and housing. India exported goods to western countries and received payments in gold. Some Roman coins were found in the southern states of India to prove this fact. Beautiful coins were also made. Art and Architecture entered a classical stage. Caves of Ajanta and Ellora were made during this period and the iron pillar which is placed near the Qutb Minar at Delhi was made by Kumar Gupta in the fifth century A.D.

Thus it becomes evident that a civilization flourishes only when there is a political and economical stability and then only a civilization can produce its creative talents in arts, architecture and allied fields. Stability and peace lead to the development of creativity and this lead to the development of the civilizations.

2.6 MANASARA : VASTUSHASTRA :

There are some elaborate treatise of town planning of ancient India. It is perhaps compiled in the 6th.C. A.D. There are several chapters on Town Planning and Building Construction and linking them, some sort of astrology. According to it there are 8 different types of towns and villages as per their shapes; they are:

DANDAKA, NANDYAVARTA, SARVATOBHADRA, PRASTARA, CHATURMUKHA, PADMAKA and SWASTIKA.

vide Fig. No. 9 — Various shapes of towns as per Vastu Shastra.

2.7 TOWN PLANNING AS PER VASTU SHASTRA :

The recommendations made in the various texts of Vastushastra (like Samarangana Sutradhara, Manasara etc.) for town planning are mainly based from the defense point of view and also taking into considerations of the various climatological and public health aspects.

Omitting the norms laid in the texts for the planning and construction of the buildings we can broadly divide the town planning recommendations into 3 basic sub-headings. These are:

(a) Regional Planning.
(b) Land and landscaping for an ideal town.
(c) Geological surveys needed for the development of towns and erection of buildings on them, along with the system of measurements.

(A) Regional Planning

Town planning presupposes the planning of the country. Shilpa Shastra recommends to have a survey of the land proposed for the development of the town. The surveys are to be conducted to check the soil condition, and the classification of the land could be done basing on the human, natural and cultural factors. The various types of land after the 'Bhu-Pariksha' or the survey for the soil condition are:

(a) JANGAL : A hot and barren land having black soil.
(b) ANUPA : The land having characteristics just opposite to that of Jangal.
(c) SADHARANA : The land having mixed characteristics of both Jangal and Anupa.

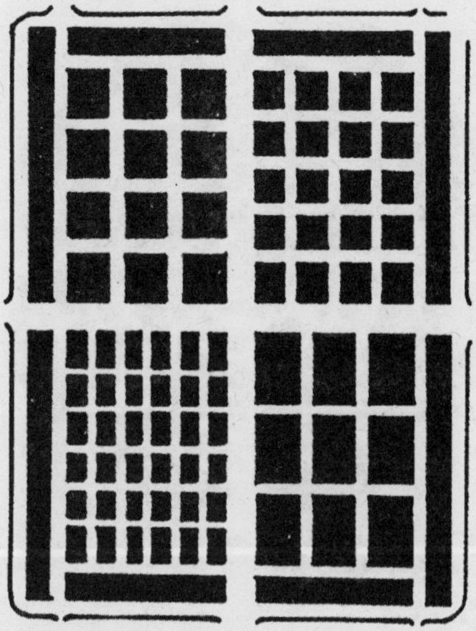

PRASTARA

KARMUKA

Fig. No. 9(a) : *Shapes of Town as per Vastushastra*

Fig. No. 9(b) : ***Shapes of Town as per Vastushastra***

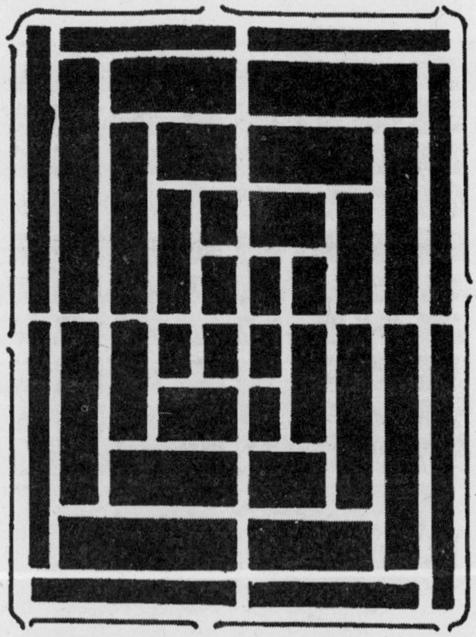

NANDYAVARTA

VARTULA

Fig. No. 9(c) : ***Shapes of Town as per Vastushastra***

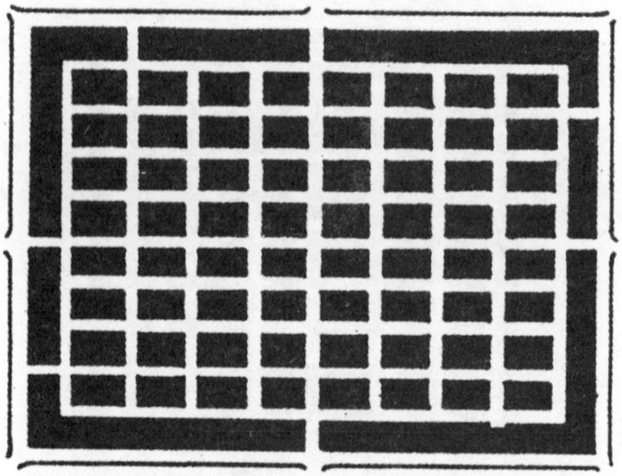

Fig. No. 9(d) : ***Shapes of Town as per Vastushastra***

Basing on the human, cultural and natural factors land could be further classified into:

(a) BALISASVAMINI: Land inhabited by good people (BHADRAJANA).
(b) BHOGYA : Land fit for enjoyment.
(c) SITAGOCHARARAKSHINI: Land having natural resources.
(d) APASRAYAVATI: Land having fearful surroundings.
(e) KANTA : Land having charming surroundings all around.
(f) KHANIMATI: Land rich with mineral resources.
(g) ATMADHARINI: Land hard to administer.
(h) VANIKPRASADHITA : Commercial centers.
(i) DRAVYASAMPANNA : Land rich in trees of superior qualities.
(j) AMRITAGHATINI: Land where people live with complete accord to each other, living in well divided 'Janapadas'.
(k) ASRENIPURUSA : Land where there is no captives and which is inhabited by gentle folks.
(1) SAKYASAMANTA : Land where Samantas are lethargic and reluctant towards state crafts.
(m) DEVAMATRIKA : Land with full of rivers and having blooming agriculture, which does not depend on rains.
(n) DHANYA : Land with easy and excessive harvests and where no ploughing is necessary.
(o) HASTIVANOPETA: Land surrounded by mountains on all sides having forests full of elephants.
(p) SURAKSHA : Land unconquerable by enemies because of natural boundaries.

(B) LAND AND LANDSCAPING FOR AN IDEAL TOWN :

(1) The land must have hills and mountains serving as natural boundaries. Where towns were made in plains, fortifications were necessary. Hence there were no much difference in the ancient manuals like MANAS ARA and MAYAMATA on the layout of towns and forts.

(2) The land must have green belts and forests all around — to support the people for food, fuel and fodder. Beauty (like Garden Cities) were also envisaged while conceiving the locations of the towns.

(3) Water sources were also necessary like the presence of rivers, lakes, ponds, tanks etc.

(C) GEOLOGICAL SURVEYS AND SYSTEM OF MEASUREMENTS :

The soils were to be tested on the basis of the five senses like the color, smell, sounds, touch and taste. SHILPA SHASTRA says that there are four types of soils suitable for the four predominant castes (or 'VARNAS, which also means'color').

(i) White was for the Brahmanas,

(ii) Red for the Kshatriyas,

(iii) Yellow for the Vysyas and

(iv) Black for the Sudras.

On the basis of taste, lands are classified again into four categories as per the four castes, namely,

(i) Sweet for Brahmanas,

(ii) Astringent for Kshatriyas,

(iii) Pungent for Vysyas and

(iv) Bitter for Sudras.

These classifications were only recommendatory and were not followed very strictly.

Different castes were placed in different parts of a city .The Brahmins and the astrologers were placed on the northern side, the market place and the traders were placed on the north-eastern corner, the location of the palaces and the royal personnel were on the eastern part, the south-east was occupied by the goldsmiths and the soldiers. The southern part was resided over by the slaves and the workers. The south-western part was occupied by the shepherds, herdsman and the fishermen. The western side was occupied by the drivers, the royal servants, the police. The municipal administrators and the merchants were settled on the north-western part of the city. Though it seems that the location of the various castes on the various parts of the city depended on the climatological conditions and the social status of the castes, all these were given a camouflage of astrology so that the common people follow them with some sort of religious austerity.

The selection of site for the construction of buildings depended on many tests, some of which is stated hereunder:-

(1) To dig a hole of certain dimensions on the proposed site. Then the hole is filled up with the earth dug out for the very same hole. Now, there can be three cases; the filled earth can form a dune over the hole, it can just fill the hole, or it can dip down at the middle. If the earth pile up in the shape of a dune, then the plot is very good for building purposes as the earth elsewhere is compact. If the earth just fills up the hole, then the soil condition is moderate and if it dips down, then the soil is unsuitable for building purposes as the earth elsewhere in the site is concluded to be loose.

(2) To dig a hole and fill the hole with water, after a night,if the water level in the hole comes down, the soil is not suitable as the soil is lose and water percolates through it, and it will hold not enough water to sustain the population.

2.7.1 Measurements:

A Human Scale was adopted for measurements in Vastu Shastra. The unit of scale was DANDA which is equal to the distance between the outstretched hands of a man. The other smaller units which followed were: JAV (grain of barley), ANGUL (finger), BALISHT (palm size) and HAST (hand).

2.7.2 Norms for selection of a building site:

The soils which are to be avoided for building purposes are:

(a) Those soils which are full of ashes, charcoal, skulls, bones, husks of corn, hairs, poison, stones or which are full of rats, anthills and gravels.

(b) Lands which are hollow, dry, zig-zag in shape, having decayed wood, subterranean pits, barren, having water current moving to the left, devoid of rains, where trees are bitter, thorny, useless, devoid of fruits, abounding in birds eating raw flesh or where worms are found.

(c) Land which is sharp and dry, too hot or too cold to touch, are to be rejected. If the soil is bitter or salty to taste, it should be avoided. Lands, when struck, that generate the sound of a jackal, camel, dog or donkey or the torrent of a crude sound like that of a breaking jar is to be avoided.

Hence the stability and the fertility of the soil is tested and ensured. Also mineral resources were tapped and building materials were explored. Land by the side of a river has a problem of floods and the proclivity of the ground is important.

Shilpa Shastra says that if the ground is slanting towards the directions presided over by Varuna, Yama, Marut, Agni or if it is depressed in the middle, it is to be avoided.

House having a low ground on the left and raised on the right is beset of many ills. Similarly if it is low on right and raised on the left it is also unfit. Further, hollowness and low ground on the back is also bad. Only the ground with west-ward slops and frontal projection is the fittest one.

In NAGAR RACHANA SHASTRA (YUKTI KALPATARU), Bhoja says that the ground should be elevated in the middle, and slanting towards east and north-east.

Practically, all the shilpashastras are unanimous in extolling an easterly declination of the ground. It offers full benefit of the morning sun.

Town planning in medieval India was probably the concern of the State, as such a wide range of perspective directions could not have been within easy reach of individual resources. Town Planning, then, was a social function.

2.8 INDIA DURING THE MEDIEVAL PERIOD :

During the medieval period, the Hindu rule came to an end and the country was divided into several independent states which never united together against any foreign aggression. From 606 A.D. to 647 A.D. King Harsha ruled over northern India when the university of Nalanda became famous. It had 1500 teachers and 8500 students. The Chinese traveler Hieu-en-Tsang visited India during this period.

From 8th. to 10th. Cent. A.D. the Rajputs became powerful over different parts of the north but they were never united. The foreign invaders took advantage of this dis-unity.

By this time the Muslim religion has taken deep roots in the mid-east. The first invasions of the Muslims were by the Arabs in Sindh. They were religious fanatics and hence their invasions were accompanied by ruthless massacres, plundering, looting and breaking of idols as well as works of art and architecture, as their philosophy was totally different from the existing philosophies and culture of the country. Nevertheless, the Arabs were impressed by the Indian scholarship, arts and culture. Indian mathematics, astronomy, music and architecture exerted considerable influence in the Middle-east countries and Europe.

Between 1011 and 1024 A.D., Mohammed of Ghazni invaded India 24 times and looted the famous temple at Somnath.

Between 1348-50 A.D., Taimur Lang invaded and he uprooted everything on his way to Delhi, destroying cities and killing lakhs of prisoners.

During the period 1200 to 1526 A .D. north India was ruled by the Sultans of seven different dynasties. They were: Ghazni, Ghori, Khalji, Tughlaq, Sayeed and Lodi. They were autocrats and most of them were cruel. The Hindus were oppressed, their properties and temples were destructed and special taxes like Jezia were imposed on them.

Allauddin Khalji invaded Daulatabad in Deccan. Thereafter, a Bahamani Kingdom of Muslims were formed at Gulbarga and Bidar. It further split up into five petty kingdoms viz. Bidar, Bijapur, Ahmadnagar, Golkonda and Berar which played important roles in the subsequent Maratha Period.

The only Hindu Kingdom in the South in the Medieval Period was that of Vijaynagar on the banks of Tungabhadra. It was established in the 14th.Century. The most prominent ruler was Krishna Deva Raya (1509-1529). Vijaynagar Town was beautiful with temples, palaces and gardens. The ruins of Hampi bear the testimony of their glory of the kingdom. The population was more than 5 lakhs and prosperous commercial activities of selling jewelry and precious stones were done. Encouragement was given to Sanskrit literature. Sayanacharya wrote commentaries on Rigveda. In 1651 Vijaynagar was lost to adjoining Bahamani Muslim Kingdom.

2.8.1 Spiritual Outlook:

Due to continuous Muslim harassment in the medieval period, the Hindus became fatalistic and spiritual in their outlook. Saints like Adi Shankaracharya preached Advaita (Non-dualism) and established the four Pithas and Shri Dyaneswara of 13th.Cent. preached Bhakti Marg, always thinking of the spiritual matters and renouncement of earthy materials. Bhakta Kabir thought of Hindu-Muslim unity and Sikhism was formed which later became a militant wing to protect the oppressed Hindu population.

2.9 INDIA DURING THE RENAISSANCE PERIOD :

The Renaissance Period of Europe corresponds with the Mughal rule in India. It started with Babar defeating Ibrahim Lodi at Panipat in 1526. The northern part of India was quickly conquered by the Mugahls but a stiff resistance was offered by the Marathas in the 17th. Cent. A.D. The Mughal Period was full of pomp and grandeur and it is marked with an architectural style of its own; which reached its peak in the reign of Shah Jehan.

During the Mughal Period, very less towns were planned and constructed. Emperor Akbar made Fatehpur at Sikri which failed, as later on it was observed that there was a keen shortage of drinking water at that place and politically the geographical location of the town was not safe enough.

The city of Jaipur was built by Maharaja Jaisingh in 1753 within 6 years after laying of its foundation stone on a carefully prepared plan with broad avenues and public buildings. Sawai Ramsingh gave the city its universal pink color and he founded all the institutions by which Jaipur became famous.

Jaipur originally had well laid roads with geometric precision and architectural accuracy with symmetrical open squares and fountains lined with uniform buildings. The original plan of Jaipur was made by Maharana Jaisingh himself assisted by Vidyadhar Bhattacharyya.

Another important happening of this period was the coming of the Europeans, particularly the Britishers into India. The British started with Bengal and ultimately established their empire after wrestling power from the Peshwas in the Deccan and the Mughal king in Delhi.

Thus it may be concluded that due to Muslim aggression and oppressions, the society turned introvert and unprogressive. Where the Hindus turned to Bhakti cult in a bid to save their religion and moral values by turning themselves 'anti-life'; the Muslims also, in their bid to keep their religious orthodoxy, cut themselves off the rest of the world. The muslims, because of their Persian ancestry, did not gave any importance to the maritime activities that was prevalent in the earlier Hindu periods. Thus, the seas surrounding the sub-continent on all the three sides were left neglected, which consequently became the key route of the foreign tradesmen to come and conquer the land.

The western world on the other hand, freed themselves from the clutches of the clergy, took full advantage of the technological inventions and made progress in all spheres of life including architecture and town planning. Hence, when the Europeans came to India, they found that here lies a country rich in resources but not equipped enough to utilize these resources for the economical prosperity of the country; which they went for utilizing to the fullest extent.

After the decline of the Mughal sovereignty the country was divided into various small independent states, which had enmity amongst themselves thus making it more easy for the foreigners to take hold of the political and economical reigns of the country.

3 THE INDUSTRIAL REVOLUTION

Basic changes that took place with the industrial revolution that took place in the 18th. Century Europe were :

- In 1765 James Watt invented the steam engine by which mechanical power became independent of hand power which lead to an increment of the overall production, trade and development of factories. Gradually there came a change in the economic leadership.
- Industrialization started factories which, like magnets, attracted workers, that demanded to have dwellings, shops and schools.
- There was also a basic change in the socio-economic scene. The society became 'Class Oriented' instead of being 'Caste Based; religion was loosing its importance. A new pattern of entrepreneurship developed along with the exploi-ation of the poor. In 1776, Adam Smith's theory on Capitalism lead to the increase of the number of workers per owner which lead to the formation of Trade Unions and the production per worker increased.
- With the increment of the need for raw materials for the industries, colonization started. Industrial entrepreneurs became politically powerful and the powers of the nobles and landlords declined.
- Geographically, areas which were not necessarily good for agriculture also started developing due to industrialization. Different patterns of settlement started comming up and Regional planning evolved as an emerging science.

3.1 DEVELOPMENT IN TRANSPORTATION :

In 1825, Steam Railroad was developed which helped in procuring raw materials and distribution of finished goods from one place to another. Mass Transit System (Public Transport) started in cities. In 1885 Electric railways started and Internal Combustion Engines were invented which also helped to increase the crowding of the streets. In 1897 Underground Railways started in Boston.

These various developments along with the development in the air transport intensified concentration in the Urban Areas.

Development in the communication systems like the invention of telegraph in 1850 and the telephone in 1876, the radio in the first quarter of the 20th. Cent.; made the world much smaller and easy communication in between cities lead to a faster growth of industrialization.

With the increase in the Urban Population, public health and safety emerged as a new science to ponder upon. Epidemics like cholera and plague and pestilence were frequent. In 1652, public water supply by gravity started in Boston. By 1820, pumping system came into existence and the methods of treatment of sewerage developed. Improper building bye-laws lead to heavy plot coverage by buildings which reduced natural drainage. Urban hygiene was nullified by congestion of buildings and intensity of landuse. In 1840 gas lighting in the streets of London started and by 1882, electricity was available for street lighting.

Thus, development of towns around factories and development of factories in towns had lead to the development of 'FACTORY TOWNS'

3.2 CHARACTERISTICS OF A FACTORY TOWN :

1. Every amenity of urban life was sacrificed to the requirements of industrial production.
2. Railroads and ships joined the factories and the waterfronts became the industrial core of the city.
3. The new industrial economy brought exploitation of the poor with poverty came the slums. New slums with row upon row of crowded workers' houses in the shadow of the factories were added to the existing slum. By 1870, many of the large cities like London, New York, Paris etc. had a density of more than 275 ppa. (persons per acre) for the city as a whole and about 550 to 750 ppa. in the residential areas.
4. No landuse standards lead to the development of buildings in a chaotic manner with no considerations for the sanitation and ventilation of the buildings.

3.3 THE CITY BEAUTIFUL MOVEMENT IN AMERICA :

A World Fair was held in Chicago in 1893 which was a reaction against the effects the factories were having on the urban life. A need was felt for a fair which would be big, broad and beautiful in contrast to the cities which were cramped, monotonous and ugly. 'Make no Little Plans'— was the move. The plans were of colossal scale with monumental proportions. It was a paradox to name the fair as a 'WHITE CITY'. The fair generated a tidal wave of City Planning. Every City Planner became a member of the 'CITY BEAUTIFUL' movement as advocated by Daniel Burnhem. Ecole des Beaux Arts in Paris acted as a fountainhead for the designers.

But these activities were detached from general public and the plans were not confirming to the human scale. Moreover with technological advancements, cities emerged as commercial centers and economics played a very important role in City Planning. Hence the need of 'practical men' and not 'dreamers' were felt necessary as cities, now, have become a business proposition.

The basic changes in the approaches of planning took place in 1900 A.D., when City Planning was looked after as an engineering process and commercialization and other tertiary activities became the most dominating reality. This resulted in the chaotic growth of cities.

Regulation for light, air, plot coverage were enforced since the beginning of the 19th. Cent, but restriction on commercial and industrial buildings started as late as the beginning of the 20th. Cent.

In India regulations on housing was incorporated in the Municipal Acts (the first Municipal Corporation Act was that of Mumbai of 1888), but regulations for non-residential buildings were only framed in 1966.

3.4 STARTING OF MODERN TRENDS IN PLANNING :

The Industrial Cities gradually turned themselves from bad to worse. The environmental deterioration was affecting the health and in turn the inefficient and unhealthy labor was affecting the production. This aroused a consciousness towards making a healthy living place for the industrial workers. In 1797, 'SOCIETY FOR BETTERING THE CONDITIONS OF POOR' was formed in England, and in 1816, Robert Owen, an industrialist, proposed a plan for a community that could be having an area of 50-100 acres, having a population of 1200 or so, which is self supporting and having a lot of open space around the community. He reformed the working conditions, wages and working hours of the workers. He set forth the unusual plan for a co-operative community combining industry and agriculture, (vide Fig. No. 10 — Concept of Robert Owen).

3.5 PUBLIC SECTOR HOUSING AND HOUSING CONCERNS :

By the end of the 19th. Cent, the governments in Europe assumed more and more responsibilities for the improvements of the city. In 1901 in New York, Tenement House Act was passed for housing reforms to improve Low Cost Housing. In 1904, Washington Sanitary Housing Company started massive housing programs.

Before that, long back in 1844, Rochdale pioneers found Consumers Co-operative Organization and in 1848 Public Health Act was passed in England.

3.6 MODEL TOWNS :

In 1890, British Housing Law empowered the State and Local authorities to condemn land and built dwelling for the working class. Many Public Utility Societies were formed in Holland, Sweden, etc. whose main aim was to raise fund, making Co-operative Housings etc. Miss Octavia Hill launched her program in London based upon the idea that good and continuous management could improve living even in the existing tenements.

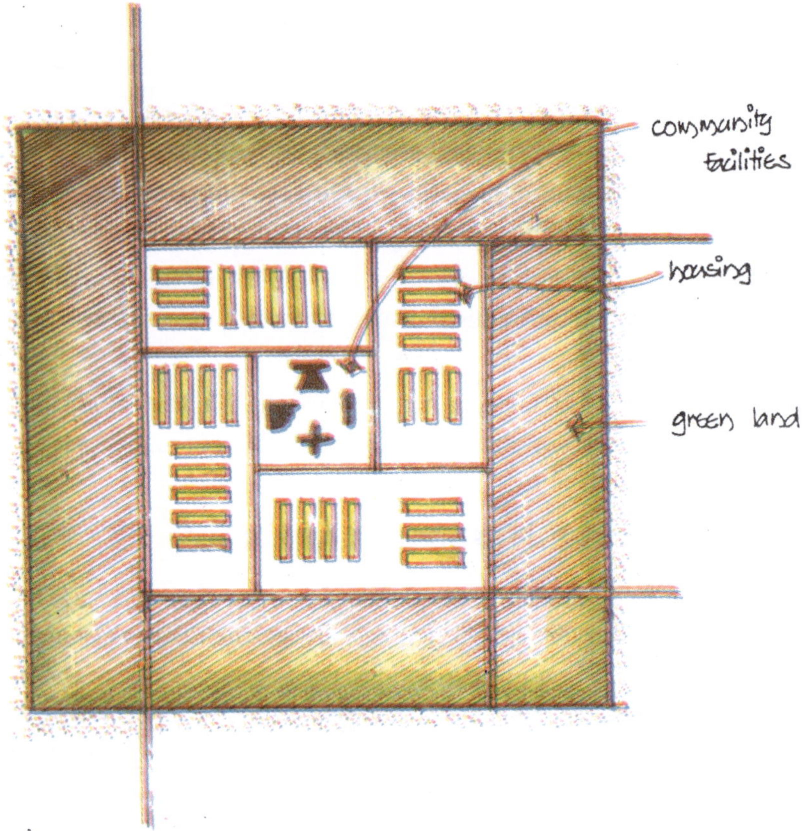

Fig. No. 10 : *Concept of Robert Owen*

To improve low cost housing the Tenement Housing Committee of the Charity Organization Society conducted a competition in 1899 which specified certain basic planning standards.

The Factory Town system brought more people to the Urban Centers and the increasing density of the tenements increased the value of land which further increased the density. It was some sort of a vicious circle.

North America had been divided into square grids by the first settlers (in 1 mile and/or 1/ 2 mile squares). As the land prices boomed, this only reinforced the square or rectangular grid planning. Natural features like rivers, mountains and valleys were ignored. Henceforth the grid became the basic pattern of forms, villages and towns, cities and countries.

The seeds of Town Planning was now planted in North America basically in two aspects, namely 'Zoning' and 'Public Utilities'.

Zoning is defined as " the regulation of the districts under the Police Power of the height, bulk and the use of the buildings, the use of land and the density of population."

Necessity of Public Utilities were felt to be imposed upon transportation via common carriers, roads, water supply, sewage disposal and drainage, communication, power and illumination.

The pyramidal form of commercial cities (from economics point of view) forced thinking on 'decentralization'. The appropriate forms of the future cities has not emerged yet, but three major areas of thought drew the attention:-

1. Planned Decentralization of New Towns.
2. To change the Internal Structure of Towns by re-development.
3. To Chalk out the Future Urban Structure.

3.6.1 Planned Decentralization :

Several philosophies for making up of New Towns.by Planned Decentralization were put forward by many thinkers. Some of which are described hereunder:

I. THE GARDEN CITY (1898)

Ebenezer Howard, a shorthand writer and a common man, published a small book 'TOMORROW focussing attention to create New Cities, to fight out the industrial hazards cropping up in the cities after the industrial revolution. The philosophy of Howard was 'To have a livable, viable and a healthy city in the Urban Area, the City should have an optimum size, a specific population and density having agricultural fields around the city.' The Garden Cities should have all the benefits of a Village and also that of a Town. It is a complete working city having a population of $\pm$ 30,000.The town area is about 100 acres. A rail road by-passes the town meeting the town at a tangent, (vide Fig. No. 11 — The Garden City).

In 1899 a Garden City Association was formed and in 1903 the first Garden City of Letchworth started, 35 miles from London.

3.6.1.1 Letchworth :

It was having an area of500 acres and was designed for 35000 persons, surrounding it there were 3000 acres of green belt. By 1947, it had about 16000 population and 100 factories. Thus the dream of Howard was shattered by the influx of people and industries and thus, gradually , there remained no difference between Letchworth and any other city.

3.6.1.2 Welwyn:

The second Garden City started in 1920 having 2400 acres and was initially made for 4000 persons. But by 1947, it was also having about 18000 population and 70 factories.

3.6.1.3 Causes of Failure of the Garden City Concept:

The concept of the Garden Cities flopped as it failed to incorporate any technological changes and the inevitable development of the cities due to these changes.

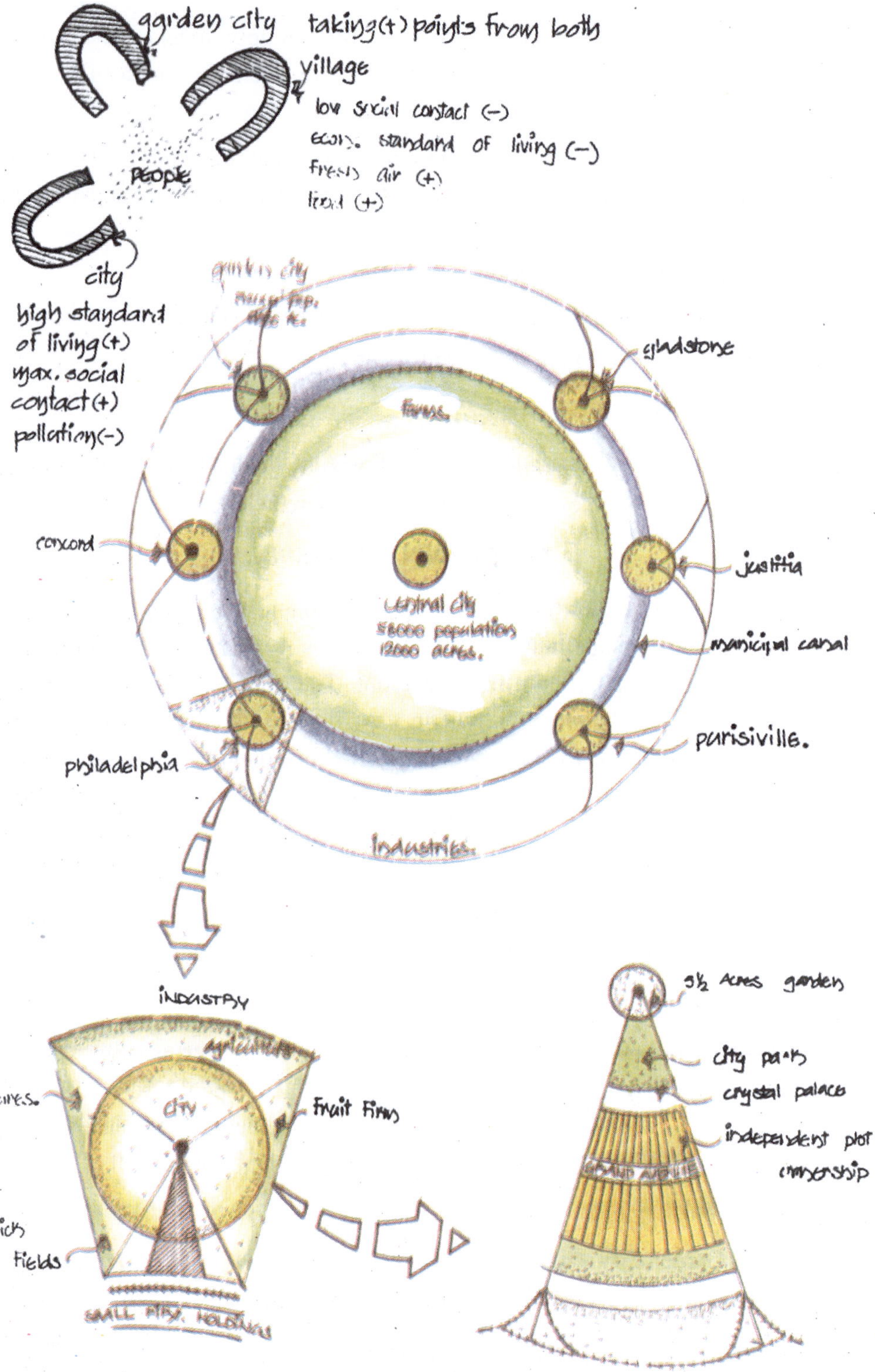

Fig. No. 11(a) : *The Garden City*

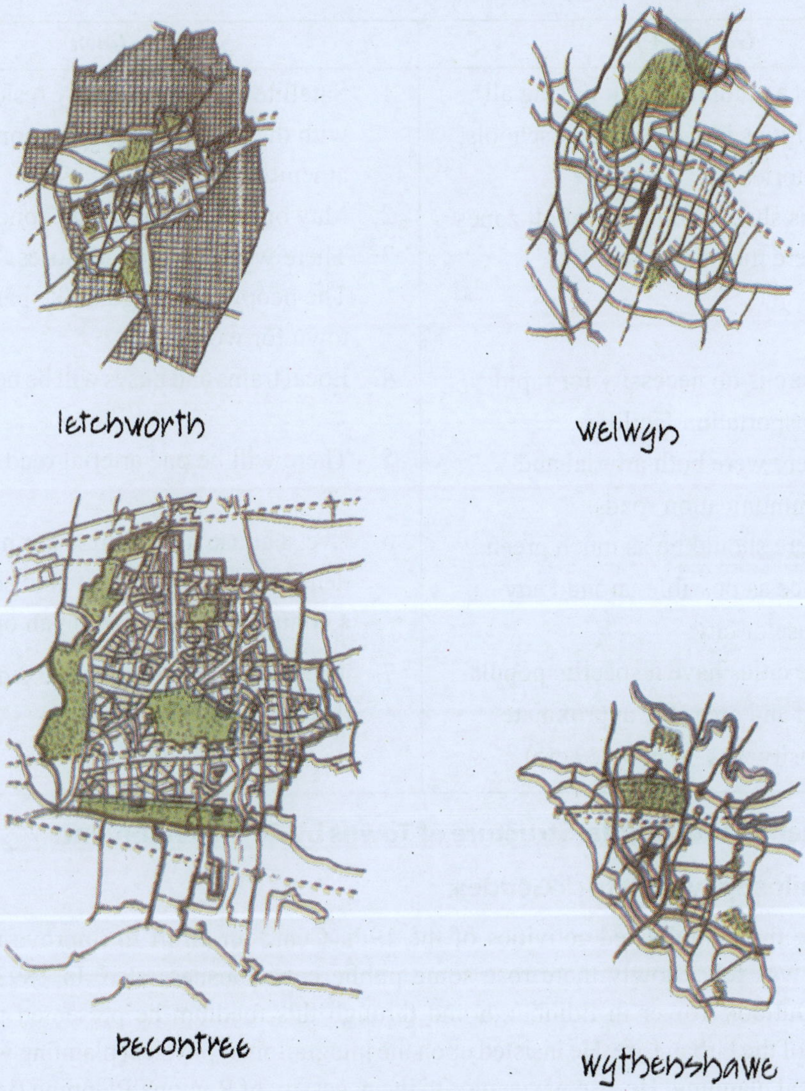

Fig. No. 11(b) : ***Garden Cities and Satellite Towns***

II. SATELLITE TOWNS

The concept of Satellite Towns was advocated by Raymond Unwin. In 1926, Manchester Housing Committee appointed Sir Barry Parker to make a 'Satellite Garden Town' to relieve the pressure due to the congested slums of the city. By 1935, Wythenshawe was built for 4000 dwellings. This was followed by the township of Becontree.

Difference between Garden city and Satellite towns

Garden City	*Satellite Town*
1. It is a complete unit; having all facilities like residences, schools, factories etc.	1. Satellite towns are mostly residential with the provision of some commercial amenities.
2. This should comprise of all zones.	2. May or may not have any zone.
3. There may be industries.	3. There would be no industries. The people would go to the parental town for work.
4. There is no necessity for rapid transportation facilities.	4. Local trains and buses will be necessary.
5. There were both arterial and communication roads.	5. There will be one arterial road.
6. There should be as much green space as possible; around any house or city.	6. Every house may or may not have gardens but the town as a whole should be a compact unit,situated in an open area.
7. The cities have a specific population and area (the approxinate density was 30 people/acre).	7. There was no concept for fixed density.

3.6.2 To Change the Internal Structure of Towns by Re-Development:

3.6.2.1 Philosophy of Patrick Geddes:

To solve the complicated activities of the 19th. Cent., an effort to improve the living environment was felt. Slowly there rose some public consciousness also. In 1892, Geddes formed the Outlook Tower in Edinburgh and through this medium he presented the whole complexities of the Urban Life. He insisted upon the integration of physical planning with social and economical planning. He also gave voice to the necessity of RegionalPlanning (in 1910) as against the German Town Extension Plan, the English Garden City and the Parisian concept of Boulevard and vista.

Geddes advocated that there is a need of simultaneous thinking in the fields of:

(a) FOLK i.e. Organism.

(b) WORK i.e. Function and

(c) PLACE i.e. Environment.

In his book 'Cities in Evolution' published in 1915, he advocated the sequence of planning to be :

(i) Regional Survey,

(ii) Rural Development,

(iii) Town Planning and

(iv) City Design.

These steps were required to be kept constantly up to date.

Geddes also advocated the theory of Conservatory Surgery. In this approach, the planners suggest that the organic growth of cities should be controlled by initiating planning bye-laws like 'zonal regulations', 'height zoning' etc., eliminating only the 'malignant' part of the structure(s). To implement this theory a method known as 'Urban Renewal' is adopted; which includes :

(a) Fixation of project area and action area.
(b) Surveys
(c) Identification of problems
(d) Making Plans and
(e) Implementation.

Geddes visited India in 1915-16 at the invitation of some of the rulers of the Princely States for rendering advise on town planning of some of their capital cities. He visited many cities in India such as Vadodara (Baroda), Varanasi (Benares), Bangalore amongst others and his recommendations were published in abook 'Patrick Geddes in India'. His visit has not, however, created any impact on the general development concepts of towns in this country.

3.6.3 To Chalk out the Future Urban Structure :

There came up certain Utopian ideas in the late 19th. and 20th. Cent. for making of the future cities of the world. The word 'Utopia' means 'perfect' but 'impractical'. With the development of the problems due to increase in population and industries of the cities, a lot of thinkers started think of an ideal city of the future. A few of those Utopian ideas are gjven hereunder:

(a) Tony Garnier — 'LA CITE INDUSTRIELLE'.
(b) Soria Y. Mata— 'LA CIUDAD LINEAR'.
(c) Frank Lloyd Wright- 'BROAD ACRE'.
(d) Ludwig Hilberseimer.
(e) Le Corbusier — 'LA VILLE CONTEMPORAINE'.
'PLAN VIOSIN'.
'LA VILLE RADIEUSE'.
'ASCORAL (C.I.A.M.).'
(f) Lewis Mumford — 'THE CULTURE OF CITIES'.
(g) C.A.Doxiadis — 'ERISTICS & DYNAPOLIS'.

(A) TONY GARNIER: LA CITE INDUSTRIELLE :

The planning was for a modern industrial city. The civic center and the residential centers were separated form the factory areas by the 'Green Belt'. The highways and the railway traverse the green belt to avoid any development abutting the regional route, (vide Fig. No. 12—La Cite Industrielle).

(B) SORIA Y. MATA: LA CIUDAD LINEAR (Linear City — 1882) :

This city was conceived as to expand along the communication network (spine) between two existing cities. Stretching along the roadway the housing and the industries bordered a continuous link between the cities. The total width of the Linear City is dependent on 'pedestrian scale' and not on 'vehicular' or 'technological scale'. SoriaY.Mata could not foresee the impact of time-distance phenomenon of the individual affecting the physical distance (like the width of the town) along with the changing transport technology, (vide Fig. No. 13 — La Ciudad Linear).

(C) FRANK LLOYD WRIGHT: BROAD ACRE (1925-30) :

Broad Acre is essentially a 'Linear City,' distributing the industries, commerce, housing, social facilities and agriculture along the rail road artery and has access to highways. One acre of land is given to each family rather than clustering them in the form of a neighborhood; though all the neighborhood facilities are provided, (vide Fig. No. 15 — Broad Acre).

(D) LUDWIG HILBERSEIMER:

In his concept Hilberseimer tried to get a relationship of the residential areas to the industrial and commercial areas. The distance between them may vary depending upon the speed of the movement. All housing were proposed in the public sector or as institutions.

(E) LE CORBUSIER (CHARLES EDOUARD JEANNERET) :

(i) LA VILLE CONTEMPORAINE (Concentric City – 1 922)

This concept was shown in Paris Exhibition. It was a city of sky-scrapers of 6 storied office buildings accommodating 1200 people per acre, and covering only 5% of the ground area. These were grouped at the City Center. The transportation center, the rail station and the air field were also kept there. Surrounding the sky-scrapers were Apartment Districts . These had 8 storied high buildings with broad open spaces around them. The density was 1200 people per acre. On the outskirts were CITE JARDINS (Garden Cities of single storied housing). The city was for 30,000 people. Most of the buildings were on stilts.

The planning philosophy was create urban surroundings as definitely contrasting to the rural areas. According to Corbusier, the densities themselves are no problem and congestion and 'slum' condition in the cities are due to excessive coverage, persistence of old street patterns and unrestricted land speculation. High density living should provide for classified street system, parking areas, adequate open areas for parks, sport fields and community services. If these are

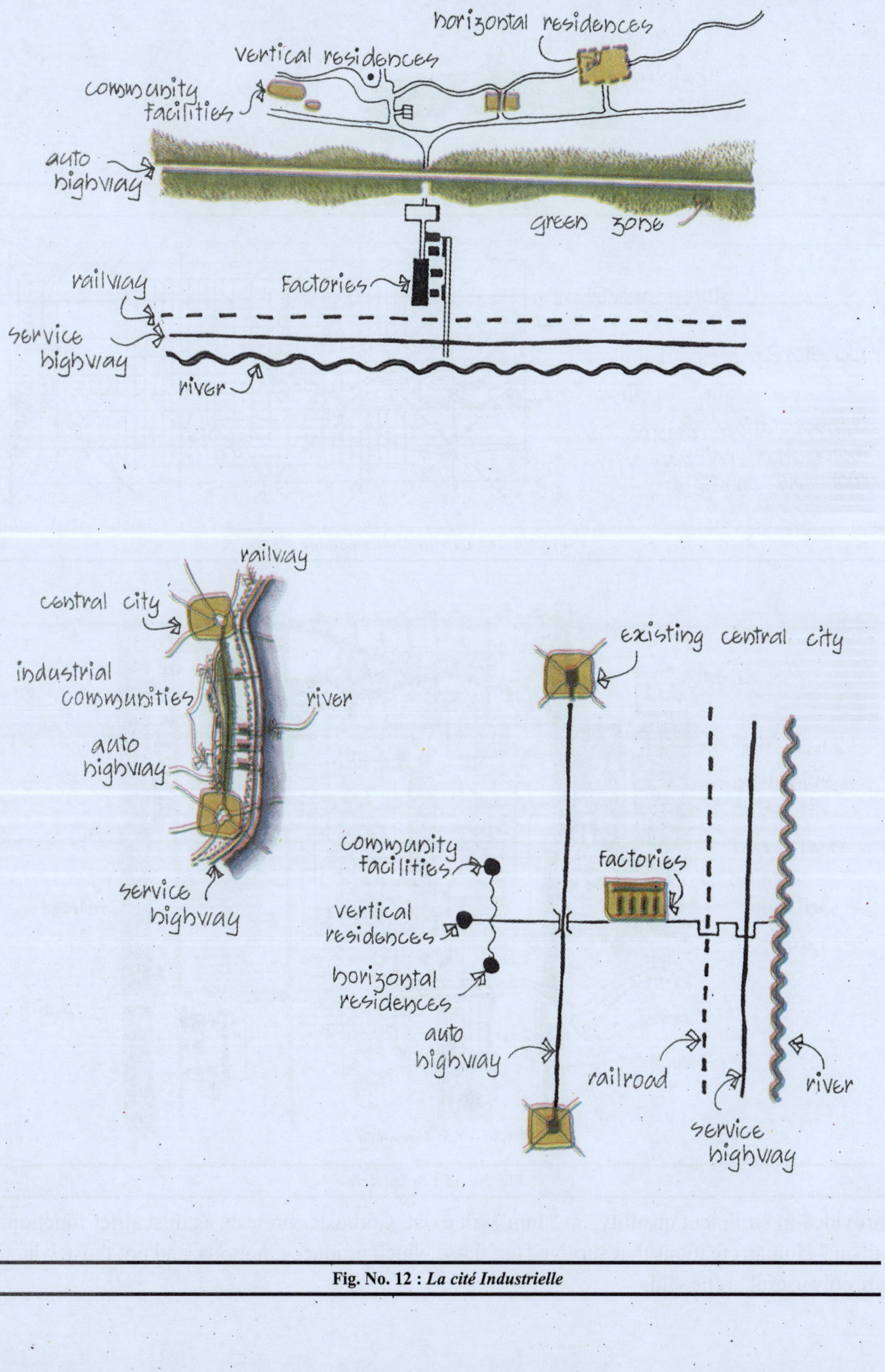

Fig. No. 12 : *La cité Industrielle*

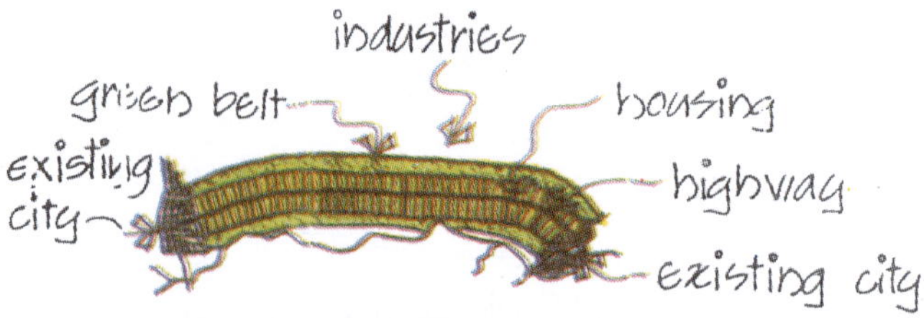

Fig. No. 13 : ***La Ciudad Linear***

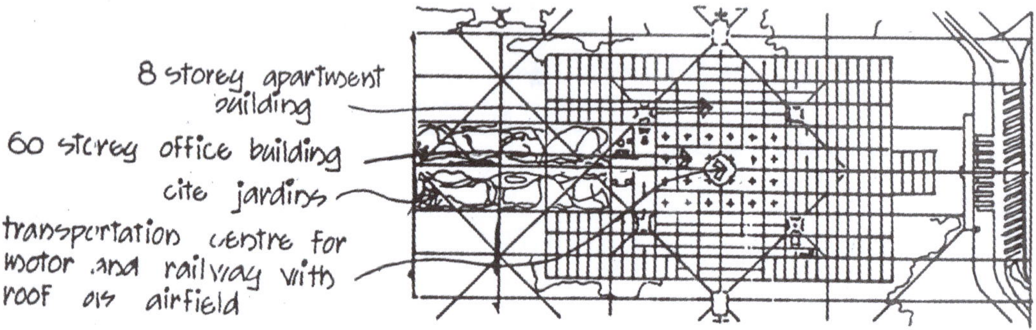

Fig. No. 14 : ***La Ville Contemporaine***

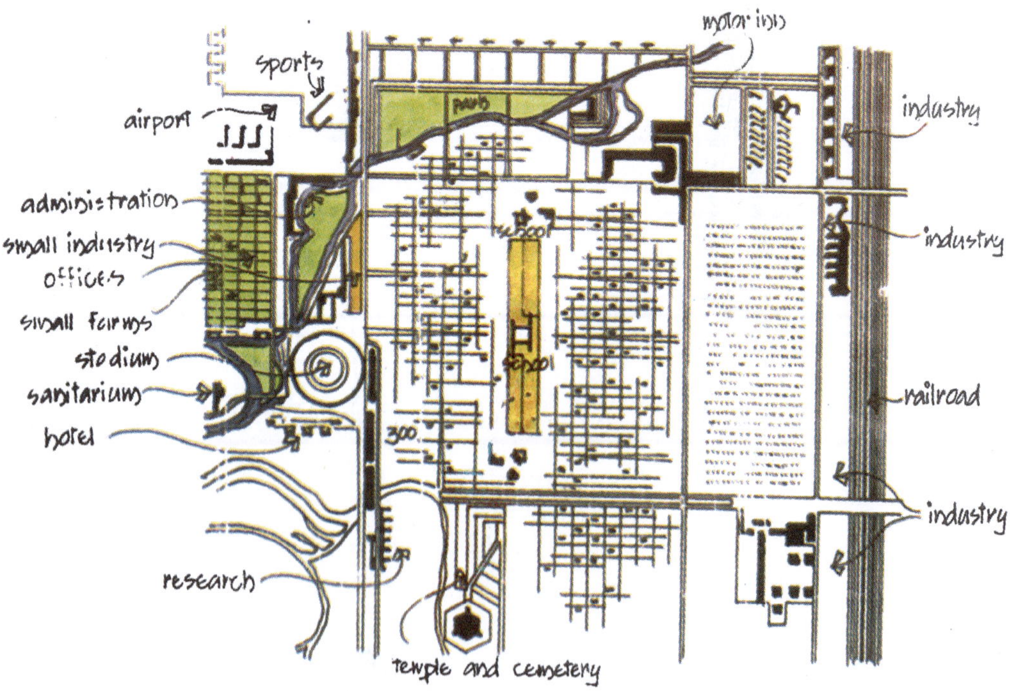

Fig. No. 15 : ***Broad Acre***

provided in sufficient quantity, no 'slum' will exist. Corbusier protests against strict functionalism;" Human creations that survived are those which produce emotions, and not those which are only useful"— he said.

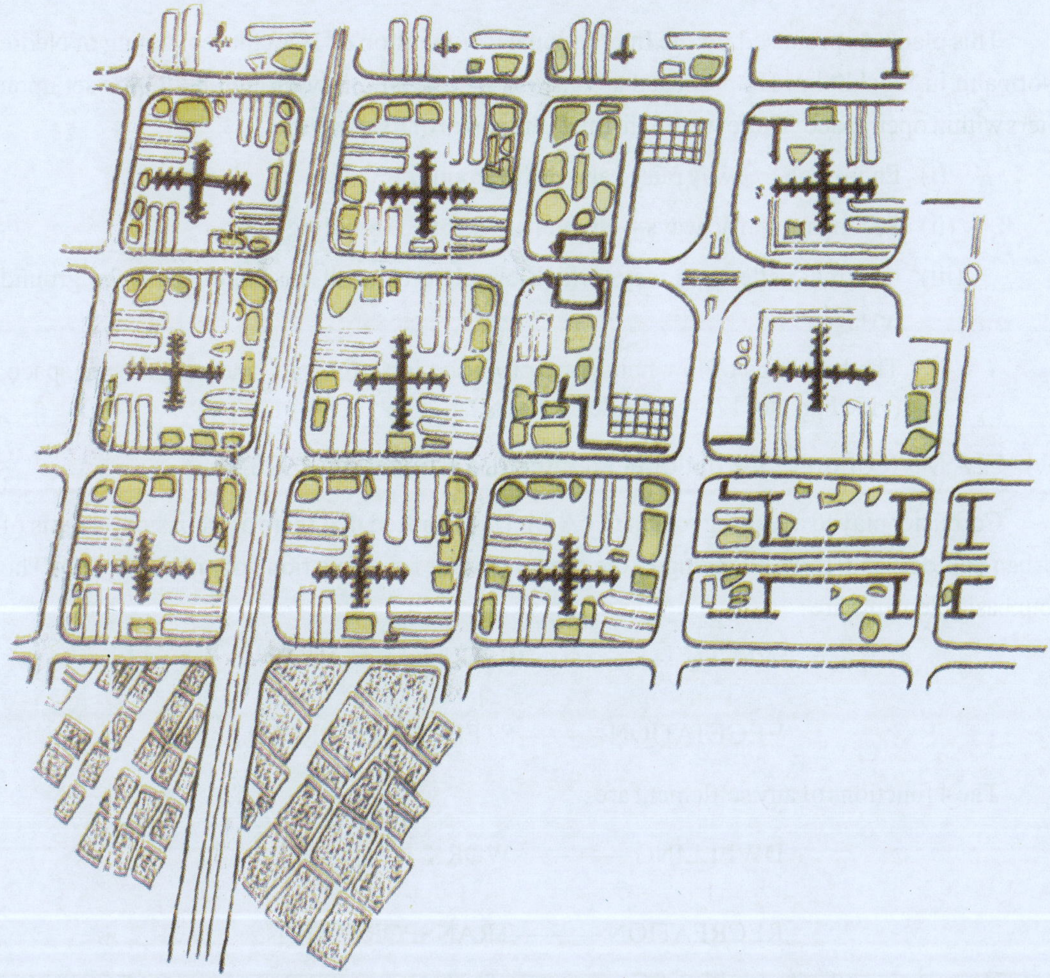

Fig. No. 16 : *Plan Viosin (1925)*

Ville Contemporaine is a revolt against the irrational growth of contemporary cities. It is a plan for a concentric city in which orderly, controlled elements replace the traditional pattern of old metropolis, (vide Fig. No. 14 — La Ville Contemporaine).

(ii) PLAN VIOSIN : 1925

The 60 story sky-scraper office buildings were placed in the vast open space, the main traffic highways were defined with complete separation of traffic and parking spaces for vehicles. The plan has rectangular arrangement of streets but local and through traffic are distinctly separated and large open spaces are treated with informal pedestrian circulation and are landscaped, (vide Fig. No. 16 — Plan Viosin).

(iii) LA VILLE RADIEUSE : (The Radiant City — 1930)

This plan was proposed for the International Competition of 1933 for replanning of Nedre Normalm in Stockholm. He proposed a continuous staggered row of high buildings set upon piers within open space. There were 3 distinct categories of roadways:

(i) Encircling freeway raised above the ground.

(ii) Secondary traffic ways— uninterrupted by the building forms.

(iii) Informal pedestrian ways which circulate beneath the buildings at the ground level.

The buildings of low height are proposed in settings of landscaped open space, (vide Fig. No. 17 — La Ville Radieuse).

(iv) C.I.A.M. : (Congres International Architecture Moderne —1928)

Corbusier played a leading role in C.I.A.M. He proposed that there are 4 basic elements of Urban Biology and Town Planning is the organization of the functions of collective life. The four basic elements are :

SPACE —————— SUN
| |
VEGETATION ——— STEEL & CONCRETE.

The 4 functions of any settlement are :

DWELLING ——— WORK
| |
RECREATION ——— TRANSPORTATION.

He organized the Assembly of Constructors for Architectural Renovation (ASCORAL); which set forth the 3 human settlements viz.:

(a) The Farming Unit.

(b) The radio-centric city (Ville Radieuse).

(c) The Linear industrial city.

ASCORAL gave importance to the basic organization of the Human Settlements, a fusion between the concentric form of the pioneers to advocate the concept of the Garden City and the Ribbon Form of the Linear City.

(F) LEWIS MUMFORD : THE CULTURE OF CITIES, 1938 :

Mumford was a student of Patrick Geddes. He was one of the pioneers to advocate the concept of Neighborhood Unit. The size of the city is determined by the convenient walking distance for children between the farthest house and the school and playground in which a major

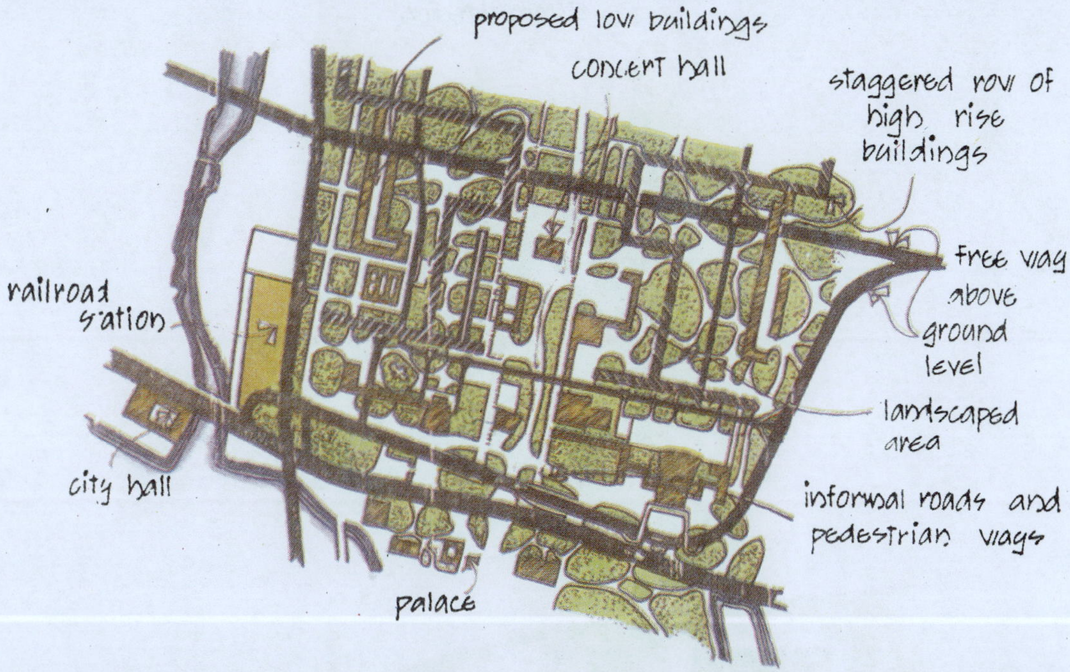

Fig. No. 17 : *La Ville Radieuse*

part of their activities are focussed. The school and the homes should be isolated from the hazards and the noise of traffic. The traffic which enters the neighborhood must be such that it subserves it directly, moving at a place that respects the pedestrians.

(G) C.A.DOXIADIS : EKISTICS & DYNAPOLIS :

Doxiadis advocated the theory of' Ekistics', or the Science of Human Settlements. He said that the Human Habitat is a balance of certain elements. These elements are:-

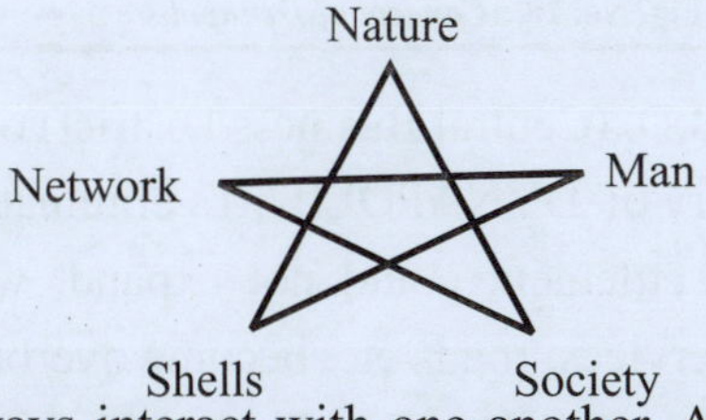

These elements always interact with one another. A human being has some invisible spheres around him. These spheres are the spheres of the senses like touch, smell, sight, hearing and also supernatural or spiritual. The spiritual sphere is directly proportional to his intellect. People interact with one another by direct interaction of these spheres. Human habitation requires a certain amount of overlapping of these spheres, and the planning of habitation would mean 'Social Planning'. Human desires and endurances have remained the same throughout the years and manifestations of which have changed by evolution.

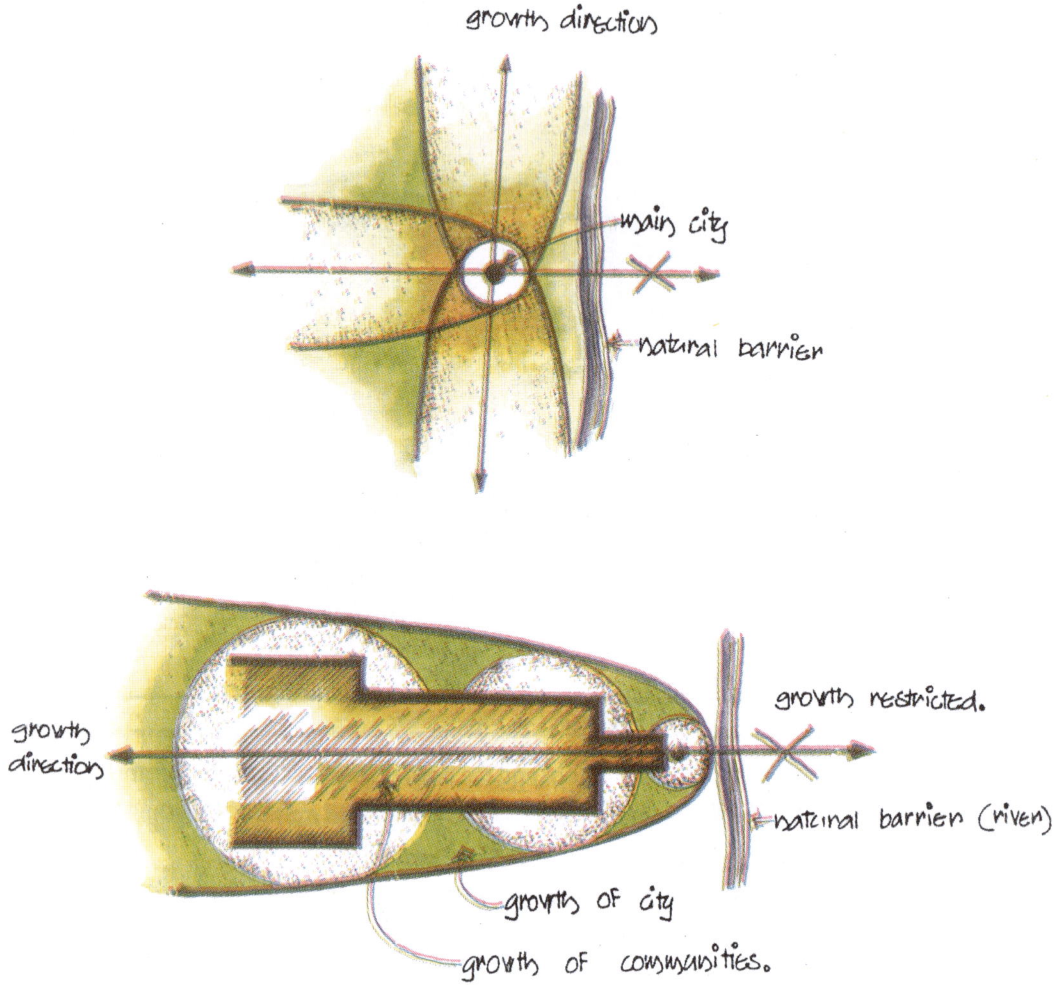

Fig. No. 18 : ***Concept of Dynapolis***

To deal with the growing C.B.D. (Central Business District) of any city and also the city itself, Doxiadis advocated the theory of 'DYNAPOLIS'. To eliminate the concentric growth of an Urban Settlement where the settlement could not expand, with increasing growth of population and consequently the services, roads etc. become overburdened. This concept is a relief to such Urban Hazards where the C.B.D. grows freely with the increase of population and such proportionate growth of C.B.D. at various phases of time cater perfectly for the ever increasing urban Settlement. The growth direction of the C.B.D. may be conceived as the principal axis of a parabola and the growth direction of the residential districts etc. is along the parabolic path. This could happen at all the sides unless stopped by some natural restraint like rivers, mountains, etc. This concept is practiced in cities like Khartoum, Islamabad etc. (vide Fig. No. 18 — Concept of Dynapolis).

3.7 EVOLUTION OF CITIES :

The city is described as a living organism. It takes birth, it grows and it dies. It also has some sort of rejuvenating power. It seems to be living for ever as we go on replacing parts of it but historically many cities (like Rome, Jerusalem etc.) died and again came back to life. Same is the case with Delhi which was made for nine time through out the course of history. There are various stages in the life of a city. These stages can be listed as follows :

3.7.1 Ecopolis:

'ECO' in Greek means 'Small'; and 'POLIS' means 'City'. This is the first stage in the development of a city. It is little more developed than village community. It shows an orderly routine of people, more security, indigenous occupations and mixing of the major works. Here association of the people takes place on the basis of blood relation and the common alter takes the shape of a community temple. Certain art forms becomes a part of life like singing, dancing etc. Granary becomes one of the important places of the ecopolis.

3.7.2 Polls:

Village agglomerate together to form a Polis. This agglomeration is done basically for defense purposes. The people here have a common God and they share the same temple. Industrial productivity rises and specialization begins to appear. Mechanization (like paved roads and wheeled traffic) are introduced and health and medical treatments are made available. Buildings begins to represent a particular way of life (like schools, gymnasium etc.). Polis also maintains links with lower order developments via migration.

3.7.3 Metropolis:

'METRO' means 'Mother'. It is a city which emerges from a group of villages. Water and food are made available and administration and machineries are introduced. The 'alphabet' of the city itself develops (like the names of streets, addresses etc.). Traders and students form two different groups in a Metropolis and hence there arises a conflict of culture as the people have different ways of life and different cultural values. This results in the development of different economic and social cultures. Centralized administrative systems also develops and agriculture becomes secondary to mechanization. Exploitation in economy gives rise to economical segregation amongst people and the gap between the poor and the rich starts widening.

3.7.4 Megalopolis:

This is the beginning of the 'end' of the city's life. The rich capitalists rule and the city concentrates on gathering wealth. Achievement is represented by monumentality in arts, architecture and social amusements. Mechanization competes with arts and architecture and there is seen an extension of 'parasitism' in economic and the social scene. Politics becomes a tool for the betterment of individuals with public funds. Growing violence amongst the people is an indicator of the slow decay of the city. There is also demands of protection money by organized thugs and 'mafiaism' develops in all the business activities.

3.7.5 Tyrannopolis:

This is a deteriorated version of the Megalopolis. Here the situation worsens and people show a trend to flow out of the city though depending on the city's infrastructural facilities for their sustenance. Thus, conurbation take place.

3.7.6 Nekropolis:

'NEKRO' means 'Dead'. There are symptoms of war, famine and disease. It shows no evidence of life and people leave out from the city in search of better place for living or try to rejuvenate it as per their own requirements.

Unit II

1 PRINCIPLES AND PROCESSES OF PLANNING

INTRODUCTION : The term 'Planning' is used in so many different ways in various literatures that often it is difficult to identify a complete and rational definition of it. It is defined by Conyers and Hills(1984) as.. ."a continuous process which involves decisions and choices,about normative ways of using available resources ,with the aim of achieving particular goals at some time in future."

Lewis Keeble (1983), describes it as 'Planning, in the broadest sense.is making up of ones' mind what to do before doing it. Town Planning is.ideally, deciding what a town should be like before it is built, though it is fortunate and exceptional for this to be possible; it is more often,deciding in what ways a town should be kept as it is and what ways it ought to be altered.'

Every new book of Planning offers at least one new variation in the definition of planning. Some of the definitions are given hereunder :

'Speaking generally, planning is deciding in advance what is to be done, that is a plan is a projected course of action'. (Newman;H.W.;1958).

'Planning is the working out in broad outline the thing that need to be done and the methods for doing them to accomplish the purpose set for the enterprise.' (Gulick,1973).

'Planning is more and more regarded as equivalent to rational social action, that is, a social process for reaching a rational decision'. (Dahl; R.A.; 1959).

'Planning.....is an activity by which man in society endeavors to gain mastery over himself and to shape his collective future by power of his reason.....Planning is nothing more than a certain manner of arriving at decisions and action, the intention of which is to promote the social good of the society undergoing rapid changes.' (Friedmann; John; 1959).

'Planning.....is that activity that concerns itself with proposals for the future, with the evaluation of alternative proposals, and with the methods by which these processes may be achieved. Planning is rational adaptive thought applied to the future and to matters over which the planners or the administrative organizations with which they are associated have some degree of control.' (Simon; Herbert A.; 1950).

'Planning is a word of many meanings. To some it means a blueprint for the future; to others it means only foresight, and action with the forward policies of the government for regulation of the economy as a whole. To some it means government responsibility to take whatever action is necessary to ensure that economic system operates efficiently, to others it means only that the government should co-relate whatever functions it undertakes towards desired overall objectives.' (Redfort; Emmette, S.; 1952).

'Planning is the process of preparing a set of decisions for action in the future, directed at achieving goals by preferable means.' (Dror; Yehezkel.; 1963).

'Planning is the application scientific method — however crude — to policy making.' (Quade; 1968).

'Planning is a process for determining appropriate future action through a sequence of choices.' (Davidoff & Reiner; 1962).

'Town Planning.... as the art and science of ordering the use of land and sitting of buildings and communication routes so as to secure the maximum practicable degree of economy, convenience and beauty.' (Keeble; 1969).

'Planning....as an attempt to formulate the principles that should guide us in creating a civilized physical background for human life.'(Thomas Sharp).

'Planning....whose main impetus is thus....foreseeing and guiding change.'(McLoughlin; 1969).

'Put another way, however, it is concerned with providing the right site, at the right time, in the right place, for the right people.' (Ratcliffe, John; 1974).

1.1 TYPES OF PLANNING :

There can be various types of 'Planning', all having the characteristics of looking ahead, making choices, and deciding the future action, where-ever possible, for attaining objectives. As per Albert Waterson (1965), these are:

(i) War-time Planning.

(ii) Post war reconstruction planning.

(iii) Town and Country planning.

(iv) Anticyclical Planning.

(v) Development planning.

A few others that can join the list may be considered as synonymous of the titles given above:

(i) Urban and Regional Planning.

(ii) Landuse planning.

(iii) Economic planning.

(iv) Social planning.

These classical titles are now being followed by some emerging titles such as :

(i) Rural planning and development.

(ii) Planning for Urban Poor.

(iii) Planning for informal sectors.

(iv) Planning for disaster prone areas.

(v) Planning for conservation, heritage and culture etc.

Some of the emerging titles may have an overlapping set of concerns or areas of operations, unless these are very carefully and explicitly defined at the initial stages.

There is no generally accepted way of classifying these different types of planning activities.Conyers and Hills (1984) have suggested four different, but related, criteria for classification.

A. The nature of planning goals.

B. **The scope of planning activities.**

C. **The spatial level of planning activity.**

D. **The operational level of planning activity.**

Waterson (1965) has followed **goal oriented classification** of planning activities (see page 56).

Scope oriented classification results into divisions of planning activity according to 'sectors', such as agricultural planning, industrial planning, transport planning, education planning and so on. This approach to planning not only suggests the subject matter of the classification but also the professional roles or position of the people involved in it.

Spatial Level of planning : Refers to Inter-national, National, Regional or Local Planning, and Settlement Level Planning.

Operational level of planning : The comprehensive National level Development Planning is subdivided into a number of more manageable operational levels.The three most common levels into which it could be subdivided are: PROJECT PLANNING, SECTORAL PLANNING and INTEGRATED AREA PLANNING.

1.2 ELEMENTS OF PLANNING :

A. To make appropriate choices.

B. To allocate resources.

C. To achieve goals.

D. Planning is for the future (i.e.it involves the TIME factor).

E. Planning is for implementation (i.e.practicable).

Conyers(1984) has pointed out that 'although planning is inevitably concerned with the future, this does not prevent planners from devoting a great deal of their attention to the studies of the past and present situations.In fact,studies of the present are essential in order to provide information about the existing conditions and the needs ;and the resources currently available for development.

While the historical studies can provide a useful basis for predicting possible future trends, planning is a complex process of decision making involving a wide range of individuals, groups and organizations. It is important that politicians, administrators,and general public should be involved in planning but the main role of professional planner is to act as a 'coordinator', collecting and analyzing the information and proposals for actions provided by others, rather than actually taking all the decisions himself.

Thus, within itself, planning involves persons from all walks of life. To make an exhaustive list of all the specialists involved in the planning process is next to impossible. A few prominent amongst them are listed hereunder:

1. Geographers.Geologists and Mining Engineers to select the site suitable for development and capable to sustain the load and cater for the needs of the future population.
2. Housing Experts, Transport Planners, Regional Planners : help in the allocation of various landuses.
3. Architects,Landscape architects, Urban designers etc.: create a congenial living atmosphere.
4. Electrical and Civil Engineers : help in providing the demands of the physical infrastructures.
5. Biologists and Climatologists: create a suitable microclimate and eco-system of the future development.
6. Economists.Social Scientists and Health Specialists also have important role in the planning process in their own fields of specializations.

Apart from these specialists, a team of Chemists, Bankers, Private Institutions, Draftsman, Business Administrators, Economists, Lawyers, Social Scientists also are a part of the planning team. Since, in a democratic setup.it is the elected politicians who represent the choices of the people at large.the overall planning process is invariably influenced by the politicians. Though too much of their involvement, vested interests, and having more administrative power than the planners themselves.sometimes do create problems in the planning process. Nevertheless, politicians and their influences should not be ignored while planning.

The product of planning is a PLAN, which may be :

A. A physical planning document that incorporates the findings or proposals resulting from planning process.

B. In the form of organizational charts.

C. In the form of annual budgets.

D. In the form of written memorandum instructing people to undertake particular tasks.

1.2.1 Relationship between Policy, Planning and Implementation :

Policy indicates the decisions about the general directions in which change or development should take place.

Planning is the process of deciding the appropriate course of actions that can bring about the desired changes or developments.

Implementation is the actual execution of these courses of action, as detailed out in 'Plan'.

Thus 'Planning'ls the intermediate stage between policy making (i.e the most coveted role of a Politician) and the implementation (i.e. undertaken by separate cadre of administrators).

It is next to impossible to determine clear boundaries between policy-making, planning and implementation or between the role of politicians, the planner and the administrator. Therefore Planning cannot be considered in isolation from the social, administrative and , in particular, political environment in which it has to operate.

1.3 TYPES OF SURVEYS :

As stated earlier, the role of a town planner is to act as a co- ordinator of the other specialists. He must be aware of the various works that the specialists do and also the scopes of their works. He specifies the data to be collected by them and he also must keep an idea of how to analyze them to get the relevant information. To collect the relevant data each specialist is required to undergo survey works and one must know the various type of surveys and the processes involved therein.

Surveys can broadly be divided into two categories depending on the area upon which they are to be conducted. They are :

(a) REGIONAL SURVEYS

(b) TOWN SURVEYS

(a) REGIONAL SURVEYS : They are those surveys which are done over a region dealing with :

(i) PHYSICAL FACTORS : Like, *TOPOGRAPHY, PHYSICALLY DIFFICULT LAND, GEOLOGY, LANDSCAPE* etc.

(ii) PHYSICAL ECONOMIC FACTORS : Like, *AGRICULTURAL VALUE OF THE LAND, MINERAL RESOURCES AND WATER GATHERING LANDS, AREAS WITH PUBLIC SERVICES, TRANSPORTATION LINKAGES* etc.

(iii) SOCIAL ECONOMIC FACTORS: Like, *AREAS OF INFLUENCE OF TOWNS AND VILLAGES, EMPLOYMENT, POPULATION CHANGES* etc.

(b) TOWN SURVEYS : They are done at much small scale and apart from the above data collected from the regional surveys it also includes:

(i) LANDUSE SURVEYS

(ii) DENSITY SURVEYS

(iii) SURVEYS FOR THE AGE AND CONDITION OF THE BUILDINGS

(iv) TRAFFIC SURVEYS and

(v) OTHER SOCIAL SURVEYS

For conducting proper survey, primarily, relevant enquiries should be framed in the form of questionnaires for presentation, when required.

1.3.1 Techniques of Surveys:

Of the various techniques of surveys that are followed, the four listed below are most prominent:

(a) SELF SURVEYS (i.e.mailing questionnaires to the persons to be surveyed).

(b) INTERVIEWS (i.e. by asking questions to the people to be surveyed).

(c) DIRECT INSPECTION (i.e when the surveyor himself inspects the situations concerned.).

(d) OBSERVERS' PARTICIPATION
(i.e when the observer himself participate in acquiring the data required.)

1.3.2 Scales for structuring questionnaires :

The questions that are asked in the questionnaires formed for doing the surveys can be of various types. Some of the asks for general things, some asks for some order of preferences or some give stress to the time interval between two incidents.Thus the **'Scales'** of the questionnaires are fixed, which can be described as follows :

(i) NOMINAL : Where there is no 'ordering', like asking of sex, age, employment in any particular service etc.

(ii) ORDINAL : Where there is a specific 'order' of choices like asking of priorities,housing conditions, climate etc.

(iii) INTERVAL : Where an interval of time is given importance like time taken to shift from LIG Housing to MIG Housing, time interval to change from two wheelers to four wheelers etc.This provides an yardstick of measurements.

1.3.3 Selection of Samples :

For conducting surveys, it is not always possible to ask each and every person about his or her opinion. Hence, certain number of persons are selected for conducting the surveys and these selected persons are known as 'Samples' of surveying. The selection of the number of samples is of utmost importance. The basic rules for selection of sample size are as follows :

(i) MORE DISASTROUS THE RESULTS OF POOR INFORMATION, LARGER SAMPLE SIZE IS REQUIRED.

That is, if the information got are poor (both qualitatively and quantitatively), the analysis done from them will be wrong. Thus, if getting incorrect results have a very disastrous effect on the framing up of the policies of planning, more number of people are to be surveyed.

(ii) THE MORE VARIED THE EXPECTED RESPONSES, LARGER SAMPLE SIZE IS REQUIRED.

That is, if it is expected that there will be various kinds of responses to aparticular question, more number of persons are to be asked, as more varied answers will help in getting different ideas of the people through the cross section of the people surveyed.

(iii) LARGER THE TOTAL POPULATION.SM ALLER THE PERCENTAGE OF THE POPULATION ARE REQUIRED TO BE SURVEYED.

That is, if the total population to be surveyed is very large then, even a small percentage of it will amount to quite a large number of samples. Depending on the time available, the money involved and many other things, the number of samples could be restricted by selecting a small percent of the total population.

The samples could be selected in various ways depending on the type of information required and the importance of the accuracy of the particular information in the survey process.The various types of selection of samples are:

(i) SIMPLE RANDOM SAMPLING (i.e. selecting samples at random without any criteria to select the samples whatsoever).

(ii) SYSTEMATIC SAMPLING (i.e. selection of the 'kth' element along a particular street, where 'k' can be any number).

(iii) STRATIFIED SAMPLING (i.e making of a homogeneous listing of the different sects of the population and selecting a certain percentage at random from each sect).

(iv) CLUSTERED SAMPLING (Where samples are selected from clusters and not from a homogeneous listing).

1.3.4 Errors in Surveying :

Getting biased, having errors in measurements, not getting any direct answer are the major errors done in surveys. Thus questionnaires should be framed in such a way so that all the answers received is properly checked. Pilot surveys should also be conducted to check forgery and bias.

1.4. THE TOWN AND ITS LANDUSES :

In India, an Urban Unit or Town is defined as that area which fulfills the following criteria:

(i) All places with a corporation,municipality, Town committee, Notified area or cantonment.

(ii) All other places which satisfy the following criteria:-

(a) Having a population of 5000 as the minimum.

(b) Density per sq. km. would be a minimum of 400; and

(c) At least 75% of the male working force should be engaged in non-agricultural pursuits.

Urban areas influence the areas surrounding them and thus the villages and other settlements might take over the characteristics of the town.Thus 'Standard Urban Areas' are defined which are having the following characteristics:

(i) They must have a core town of population more than 50,000 (as per 1971 census).

(ii) They must be joined with other urban and rural administrative units which have close mutual socio-economical links with the core town.

(iii) There is a probability that these areas will get fully urbanized within 2 to 3 decades.

Basing on the above definitions the cities are classified into various 'classes' depending on their populations.

CLASS I: Having population of more than 100,000.

CLASS II: Having population between 99,999-50,000.

CLASS III: Having population between 49,999-20,000.

CLASS IV : Having population between 19,999-10,000.

CLASS V : Having population between 9,999-5,000.

CLASS VI: Having population less than 5,000.

Cities having population more than 1,000,000 are known as Metropolitan Cities.

As per 1981 census, in India there were 12 Metropolitan cities namely, Calcutta (9.17m pop.); Mumbai (8.23m pop.); Delhi (5.71m pop.); Chennai (4,28m pop.); Bangalore (2.91m pop.); Hyderabad (2.53m pop.); Ahmedabad (2.5lm pop.); Kanpur (1.69m pop.); Pune (1.69m pop.); Nagpur (1.3m pop.); Lucknow (1.01m pop.) and Jaipur (1.00m pop.). The other major towns on the verge of becoming Metros were: Coimbatore, Patna, Surat, Madurai, Indore, Varanasi, Agra, Jabalpur and Vadodara. These are followed closely by Dhanbad, Bhopal and Ulhasnagar.

By 1991, the scenario have changed considerably and a look at the Class -I towns with their-population will make the picture clear.

Mumbai (125.7 lakhs pop.); Calcutta (108.6 lakhs pop.); Delhi (94.2 lakhs pop.); Chennai (53.6 lakhs pop.); Hyderabad(42.7 lakhs pop.); Bangalore (41.6 lakhs pop.); Ahmedabad (33 lakhs pop.); Pune (25 lakhs pop.); Kanpur (21 lakhs pop.); Nagpur (17 lakhs pop.); Lucknow (16.5 lakhs pop.); Surat (15 lakhs pop.); Jaipur (15 lakhs pop.); Cochin (11.5 lakhs pop.); Coimbatore (11.35 lakhs pop.); Vadodara (11.15 lakhs pop.); Indore (11.04 lakhs pop.); Patna (10.99 lakhs pop.); Madurai (10.94 lakhs pop.); Bhopal (10.64 lakhs pop.); Vishakhapatnam (10.50 lakhs pop.); Varanasi (10.27 lakhs pop.) and Ludhiana (10.12 lakhs pop.).

[NOTE : 1 lakh = 100,000]

A look at the total urban population in India from 1951 to 1991 will make the scene more clear :

Census year	Urban population (in million)
1951	62
1961	79
1971	109
1981	160
1991	218

The total area of town can broadly be divided into Undeveloped and Developed areas. Undeveloped areas are those which come within the boundaries of the Local Government but are used either as agricultural land, green belts or are undevelopable due to the presence of water bodies, marshy lands, hills etc.

The Developed area of the town may be placed under the following uses known as the Landuses of the town. The various landuses, their approximate percentage of total developed area and the color codes followed in India to designate these landuses on the Master Plans (i.e the British System of color coding) are listed hereunder :

[It is to be noted here that the percentages given are only approximate estimations and these vary from one city to another depending upon the character of the city, the economical condition and various other factors.]

Table 1.4.1

Landuse	% Age of Dev. Area	Color code
Residential	50	Yellow
Commercial	02	Red
Industiral	10	Violet
Public / Semipublic	12	Blue
Recreational	05	Green
Transportational	20	Black
Vacant Land / Incidental Open Space	01	None

1.5 THE CHARACTER OF A TOWN :

The character of a town depends mainly on the' Functional Category' of the town, or simply, the occupation of the people. In India, the occupations are divided into 4 categories such as :

(I) PRIMARY : Including (i) Cultivators; (ii) Agricultural labors; (iii) Livestock rearing, fishing, foresting, hunting, plantation, orchard keeping and allied activities; (iv) Mining and Quarrying.

(II) SECONDARY : Involved in handling of goods either to change the form or to change the location. This includes, (v) Manufacturing and processing, Servicing and repairs. This can be again divided into (a) Household industries and (b) Other than Household industries. (vi)Construction.

(In inter-national level, transportation is also included in this category).

(III) TERTIARY: Involved in performance of service or to transfer the ownership of products. This include, (vii) Trade and Commerce, (viii) Transport, Storage and Commerce, (ix) Other Services.

(IV) QUATERNARY:Involved in keeping of records.

Basically, there are 5 categories,namely, Primary, Secondary, Trade and Commerce, Transport and Communication and Other Services.

1.6 CATEGORIES OF A TOWN :

Basing on the above mentioned functional categories, towns can be classified as :

(i) **Monofunctional Town**: If 40% or more of the total workers are in one occupation.

(ii) **Bifunctional Town** : If the percentage is less than 40%, then the next predominant activity is taken and the total of these two must be 60% or more of the total workers.

(iii) **Multifunctional Town** : If the total made above, does not come to 60%.

1.7 DENSITIES OF A TOWN :

To understand (both quantitatively and qualitatively) the pressure and the future requirement of the various landuses of a town, densities are the most important tool in the hands of the planners. There are various types of densities used for various purposes. Some of which are stated hereunder.

$$\text{Overall town density} = \frac{\text{Total population}}{\text{Total town area (including the undeveloped area).}}$$

$$\text{Developed area density} = \frac{\text{Total population}}{\text{Total developed area}}$$

$$\text{Gross residential area density} = \frac{\text{Total population}}{\text{Total residential area}}$$

$$\text{Net residential area density} = \frac{\text{Total population}}{\text{Net residential area}}$$

NOTE : NET RESIDENTIAL AREA is defined as the land included in house plots, residences, roads dnd incidental open spaces only.Generally, for every one hectare of residential land,two hectares of non-residential land is required.

The densities in 'persons per hectare' will not give the true picture of the scenario as the number of dwelling units per hectare might vary. Hence we also use,

$$\text{Accommodation density} = \frac{\text{Number of habitable rooms}}{\text{Area}}$$

$$\text{Occupancy rate} = \frac{\text{Number of persons}}{\text{Number of habitable rooms}}$$

$$= \frac{\text{Number of persons / area}}{\text{Number of habitable rooms / area}}$$

$$= \frac{\text{Population density}}{\text{Accommodation density}}$$

The ideal occupancy rate is 0.8, e.g., 4 persons per 5 rooms.

In non-residential areas, population densities are not applicable and as such the measurements are taken as F.S.I.; F.A.R.; PLOT RATIO etc.

$$\text{Floor space index (F.S.I) (British)} = \frac{\text{Total floor area}}{\text{Total plot area}}$$

(F.S.I, is expressed in absolute numbers like, 1,0.75 etc.)

$$\text{Floor area ratio (F.A.R) (American)} = \frac{\text{Floor area on all floors}}{\text{Plot area + 50\% of the area of the adjacent road which have width of more than 27m. (80' 0")}} \times 100$$

(F.A.R. is expressed in percentages like, 100%, 75% etc.)

This creates problems at the corner plots as the roads to be taken into consideration overlap each other and also more floor area generated at a comer plot of non-residential use generating more traffic and more pressure on the other physical infrastructures.Hence PLOT RATIOS are used for corner plots in America.

$$\text{Plot ratio} = \frac{\text{Floor area in all floors}}{\text{Total plot area}}$$

For knowing the number of dependent population on the workers WORKERS' PARTICIPATION RATE is used.This is defined by :

$$W.P.R. = \frac{\text{Number of workers}}{\text{Total population}} \times 100$$

In India, the W.P.R. as per 1971 census was 33%.The average house-hold size was 4.8.

1.8 ZONING :

The concept of Zoning is an American innovation which came as a reaction against the haphazard growth of sky-scrapers.New York (in 1961) issued "Zoning Ordinance" for the first time even before the Master Plans were made. In India,the Town Planning Regulation does not specifically say about zoning. But zoning was practiced in the Delhi Master Plan (1962);and in the Maharashtra Regional and Town Planning Act (1966).In 1978, the Model Landuse and Zoning Regulations were formulated by T.C.P.O.(Town and Country Planning Organization.) which accommodated the concept of "mixed landuse'.

1.8.1 Concept of Zoning :

The basic concept of zoning lies in the division of a settlement into zones and districts according to present and potential uses of land and buildings. Zoning has to pay particular attention to the character of a place and its suitability .Thus, 'zoning 'is establishing 'Districts' within which certain rights of citizens are legally curbed.

1.8.2 Objectives of Zoning :

(i) To lessen congestion in streets.

(ii) To secure safety of fire and other hazards.

(iii) To promote health and general welfare.

(iv) To protect amenity value and hence property.

(v) To provide adequate light and air.

(vi) To prevent over-concentration of population.

(vii) To facilitate transportation, water supply, sewerage, schools, parks, etc.

(viii) To encourage the most appropriate use of land ;—giving the character of the districts.

Thus Zoning Regulation includes provisions for the use of property and exercise limitation upon the shape and bulk of buildings that occupy the land .The most common type of zoning are the Use Zoning and the Height Zoning.

1.8.3 Differences of Building Bye-Laws from Zoning Bye-Laws:

Building Bye-Laws	Zoning Bye-Laws
(i) Same for whole city.	(i) Different in different districts.
(ii) Deal with internal part of a building	(ii) Deal with the whole district.
(iii) They are not concerned with the use of buildings.	(iii) They are concerned with the use
(iv) Need not have a Master Plan.	(iv) A Master Plan is necessary.

1.8.4 The common aspects of the Zoning and the Building Bye-Laws are:

Both give the regulations to control height, set-backs, plot area and F.S.I

1.8.5 Pre-rcquisites for Zoning Bye-laws:

(i) The city must have a Master Plan.

(ii) It translate the proposals of Master Plans into practice.

(iii) It should be directed towards the protection of public health, public welfare and safety.

(iv) No discrimination or capricious intent should be there in the plan.

(v) Administration of the ordinance should be reasonable and free from arbitrary decisions.

(vi) Detailed studies of population , economics and landuse studies are required.

(vii) Should be adopted all over the planning area.

(viii) Efficient planning and enforcement machinery is required.

1.8.6 Types and Classification of the Uses in a Town :

(i) PERMITABLE : Uses which are automatically permitted.

(ii) PERMISSIBLE:Uses permitted on appeal.

(iii) PROHIBITED/NON-CONFIRMING:Uses not permitted.

Any use of structure, land or plot area which does not confirm to the use, height, bulk or plot area requirements of zoning regulations is 'non-confirming'. If these uses are already existing at the time of introduction of the zoning regulations, the non-confirming uses are allowed to continue under certain conditions. These are:

(i) They cannot be enlarged, reconstructed or changed from one non-confirming to another non-confirming use.

(ii) Non-confirming uses cannot be made if the property is destroyed (in time).

(iii) Unless it is a clear public nuisance, it is to be allowed to continue.

(iv) It can be eliminated by 'eminent domain'—bulk acquisition.

(v) It can be eliminated by amortization (i.e. after amortization period, the owner has to eliminate it.).

1.8.7 Zoning Aspects:

Amendments in the zoning regulations should be done for :

(i) Reflecting the character of a particular district and the bye-laws should be framed for public interest.

(ii) Spot Zoning may be done, which is, relaxation of zoning bye-laws with reference to a particular site. (i.e. location of a petrol pump in a residential area.).

(iii) Variances can be done in zoning bye-laws.In the Western countries, they are governed by a Board of Appeal. But, if any bye-law is adverse, it should be amended instead of giving variances to a number of plots or buildings.

1.8.8 Major classification of Zoning Districts:

Open land district: Given to scenic or historical sites having no urbanization.
Agricultural district:

Estate district: Given to residential districts to a large plot. **Single family district**: Given to areas having single dwelling units per plot.

Multiple family district : Given to plots having two families, with medium density developments, high density developments, mobile houses, hotels etc.

Commercial district:

Historical preservation district:

Industrial district:

Districts might also have restriction on height and bulk. Zoning should be made flexible to accommodate any change in future. In developed countries, new districts of zoning are also felt necessary like districts for car parking etc.

1.8.9 Drawbacks of Zoning Regulations:

(i) Practice is cumbersome.

(ii) Not useful for smaller towns.

(iii) Zoning bye-laws compromises with existing use, so very rarely they guide new developments.

(iv) Zoning emphasizes the distinction of the various uses and not the relationship of them.

(v) It fails to prevent change of use and coming up of uses like temples, schools etc.in public open space.

(vi) Non-confirming uses must be limited in bulk to properly regulate the environmental degradation.

(vii) Zoning laws seems to be very rigid for India as we are accustomed to mixed landuse, both at horizontal and vertical level. A detailed research is felt necessary as mixed landuse is depend on the regional culture.

(viii) In America, the major criticism of zoning is that it leads to social and economic segregation.

2

PLANNING THEORY

2.1 PLANNING PROCESS:

As we have seen in articles numbers 1.1 and 1.2, there are various types of planning for various situations and hence there are a number of probabilities to be dealt with. Along with it, there are also a number of alternative courses of actions which may take place due to various social, economic and physical constraints. Modernization may come because of changes in the political situations and the changes in the availability of resources. The Physical Planner has the role of a co-ordinator. The Planning process is a multifaceted process where various aspects are simultaneously considered. The whole process is a dynamic process as it changes with time, and it makes physical planning 'a continuous process'. Thus the system is always in a state of flux— it is getting defined and redefined with every new piece of information coming in its way. Even before 50 or 55 years Planning Process was considered to be a Linear Process. This process is also known as 'Conventional Planning Process'. The steps of this process may be stated as :

1. GOAL FORMATION (or PROBLEM IDENTIFICATION)
2. SURVEY AND INVESTIGATION
3. ANALYSIS
4. DEVELOPMENT/IDENTIFICATION OF ALTERNATIVES (these alternatives are made on the resources available and the time allotted for the project).
5. SELECTION OF ALTERNATIVES (this is a process of selection by elimination and the selection is again guided by resource availability and time)
6. IMPLEMENTATION
7. MONITORING AND FEEDBACK

Planning is not a universal package; it must recognize variety. This variety is nothing but 'Entropy'— having endless variety. Thus to bring in this entropy in some kind of manageable sequence is also a function of the planning process.

If any modification is needed after monitoring, it is fed back again to goal formation. But, the various stages of the said process have to function simultaneously, hence, the process cannot be a Linear one.

2.2 MODIFICATIONS OF THE PROCESS :

Physical Planning results in the formation of Plans which perpetuates a sense of control over development. 'Control' is a negative action on unauthorized developments. This negative action must have certain positive promotion aspects in the plan. The negative action (i.e the 'don'ts' in the control mechanism) ultimately alienates the small investors whereas the positive actions (i.e the 'do's of the control mechanism) and promotional aspects leads to the participation of the investors.

Chadwick (1971,1978) have put forward the **'System View of Planning.** The steps of which may be stated as :

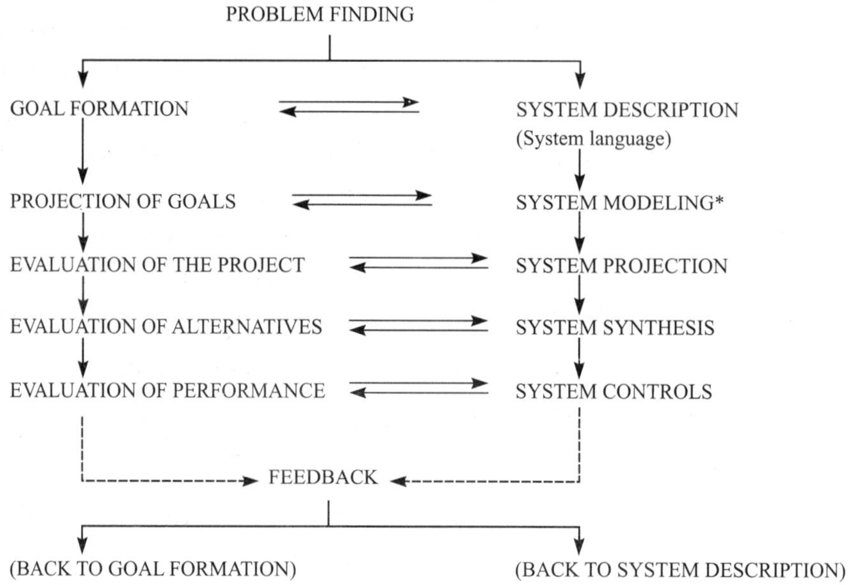

[*. By knowing the system and the attributes, we are to present it in a logical sequence, therefore, we put in the logico-mathematical expressions to explain the relationships with the elements and attributes].

2.3 APPROACHES TO PHYSICAL PLANNING :

Planning is sometimes described as a process to allocate chunks of scarce resources in a logical manner for the betterment of the society. Thus resources (which are scarce) must be employed in such a way that maximum return is got. Hence there are two approaches to Planning. One is Rationalistic Approach, and the other is Incremental Approach.

In Rationalistic Approach we go in for a Comprehensive Plan; but the cost (in terms of money and time consumption) of Comprehensive Surveys and Comprehensive Plan in much more.

The other approach is Incremental Approach, where plans are made in terms of small projects; the cost of surveying and plans of which are much less. This also get the support of the politicians as it shows quick results to the people. But, the main drawback of this approach is that, each project may not add up to a Perspective Plan on its own unless it has a Perspective Plan from the very beginning.

None of the small project is a complete solution to the problems, hence we must have an integrated system.

2.4 GOALS AND INDICATORS OF PHYSICAL PLANNING :

Planning process aims at certain dimensions of change — it aims to increase the access of services and facilities. Thus, 'Development' is a concept of 'Growth + Change'. A planner must concern himself with both of these criteria.

The 'change' is defined as 'clear identification of aims'; and thus, change is defined in terms of Goals. Goal has a 'Time' aspect with it. To achieve a Goal within a time span we must understand the nature of the statement (of the Goal). The statements are motivated to the 'social welfare'. Any Goal statement is a general statement which defines the purpose of planning; and the statement cannot be arbitrary as it has to be implemented and thus the statements must be 'practical' in flavor. To make it practical, we must also consider certain constraints such a material constraint, money constraint, time constraint etc.

Each family has a set of life which cannot be ignored while setting out the goals. Social welfare is commonly acceptable to everybody, but the tangible constraints and the intangible constraints also influence the formation of goals. The tangible constraints does not change but the intangible constraints vary from place to place and from time to time. Thus 'Objectives' are formed, which are time bound and may change in time. Physical planning, thus, cement the 'Goal statements'(which are general in nature) with the 'Objectives'(which are specific).

The Goal statements may change in certain time, as priorities may change in time. So, re-ordering of priorities may become necessary. Hence Goals are divided into long term and short term goals. The long term goals are amenable to Rationalistic Approach of Planning and the short term goals are amenable to the Incremental Approach.

The purpose of planning is to define the change and to create the means to bring in the change. To see whether a plan is correct or not we adopt monitoring, i.e., we device some kind of tests to indicate to us whether the plan is approaching the proper direction or not. These are the 'Indicators'.

Churchman (1968) said that the user's reactions has not been recorded properly over the world. So, it does not become a kind of feedback of the planning process. To ascertain whether the planning process is aiming at proper direction or not, there is very little experience to bank upon. So, building up of 'Indicators' become necessary.

Urban Indicators are supposed to define the stages of development. A planner is concerned with the living, working, recreation and other activities of the people and hence, three types of Indicators are identified under Urban Indicators. These are :

1. **SOCIAL INDICATORS :**

These are general indicators which can be broadly defined as 'measures of aggregate wellbeing among the persons and families within a society, a city, a neighbourhood.' (Stuart, 1976)

2. **IMPACT INDICATORS :**

These are specific indicators to asses the impact of a plan. Following a plan, the items which has been provided are closely scrutinized to see whether the promises made in the plan are fulfilled or not.

3. **PERFORMANCE INDICATORS :**

These are precise indicators. When various activities of a town starts functioning, then we realize whether the plan is functioning or not as many frictions amongst the different items may arise. This can be got by 'Cost-Benefit' analysis. The plans should not only be economically viable and physically possible but also be socially acceptable. Performance Indicators are calculating, so they are precise.

The goals and the indicators has a very sensitive relationship amongst themselves. Physical planning can be de-linked from the National Economic development Plan but the exercise of goal formation is not local, as the political and economic climate influence the formation of them.

2.5 STEPS OF GOAL FORMATION :

The following are the steps to formulate a Goal as advocated by R.C.Young (1964):

1. Establish parameters of concern.
2. Establish range of choice.
3. Establish the relationship between the Goals.
4. Relative evaluation of the Goals.
5. Statements of Goals as Policies.

2.6. COMPONENTS OF INTERNAL STRUCTURES OF CITIES AND TOWNS :

Structure defines the dimension, characteristic elements, arrangement of elements, and the relationship between them. It also tells about the reasons of why it behaves the way it behaves.Thus, there are certain 'causes' for the structures.

Internal structure of a city is given by the natural features, landuses, circulation areas, service networks, and explaining their arrangement, i.e., the processes which have arranged them in a particular manner. These processes are social, economical and historical background of the community for which planning is made.

The internal structure of a city is different from Landuse Planning as Landuse Planning does not gives the reason of its arrangement in space, landuse is only the relationship between a parcel of land and the human activities on it.

To find out the components of the internal structure of a city, there are certain preconditions. There must be a 'hypothesis' which has to be tested and then 'theories' are made. Theories are of two types, namely,

1. NORMATIVE : Which tells about the things to be done; and
2. POSITIVE : Which tells us how to do it.

Sometimes the Theories remain only at the level of postulates — as a set of observations which have not found enough testing ground, but within certain preconditions, they are right. The purpose of Normative Theory is to identify decisions areas in planning. The Positive Theory tells us how to behave in these decided areas.

2.7 LANDUSES IN A CITY :

Landuse is the 'Use' of the land of a city. Landuse Plans show us the various kinds of activities that are carried out in the different location of the city. There are various types of landuses in a city. **A** detailed classification of which may be seen in article **1.4;** and in Chapter III.

2.7.1 Key Systems Affecting Landuse Plan :

There are three major systems which affect the landuse plan of a city. These are:-

1. ENVIRONMENT SYSTEM
2. **ACTIVITY** SYSTEM
3. LAND DEVELOPMENT SYSTEM

1. ENVIRONMENT SYSTEM

The environmental systems are of two types namely,

A. Biotic and

B. Abiotic.

A planner modifies the natural system existing on a parcel of land with the help of certain 'moderators'.

The Abiotic factors are soil, air and water and the Biotic factors are the flora and fauna, which are to be altered to the extent that is necessary for the human activities. The flora is related to the temperature and rainfall. As the greenery is cut down, the micro-climate changes and hence rainfall changes. Planting trees does not necessarily bring back the existing micro-climate. These trees are'domesticated trees', which may not belong to the eco-system. So, the number of birds etc. diminish and hence, the characteristics of the local eco-system changes.

Natural water bodies act like 'rechargers' of the underground water table. If ponds and natural water bodies are covered up, the water table may go down. So, after a certain time there may be water shortage at different places of the city. If too much water is pumped out from the water table, and if no re-charging is done, then, the increasing weight on the soil due to building activities, lead to a sinking of certain pockets of the city. Cracks start developing on the buildings due to subsidence. This happened in Mexico. The city of Calcutta is also sinking at the rate of 75 cms. a year.

Cutting of forests loosens the top soil, and flash flood takes place. The loosened soil are washed out and they go the water bodies and deposit at the bottom; thus decreasing the depth, which again, becomes, the cause of flash floods.

Thus, hydrological data (showing the quantum of water available and how much of it is replaced) together with geological data (showing the soil and the rock structures) are very important things to be considered before selecting a parcel of land for any activity.

2. ACTIVITY SYSTEM

The Activity system can be best understood if we list out the various agents that perform the various activities on land. The list given below, will make it clear.

(i) Individuals/households [individuals have preferences based on social background, economic status and education]	Living, working, recreation & communication. [concept of affordability becomes a check, so, depending on affordability, location of residential areas are to be distributed. The location of residential areas is a function of distance to the town center to schools, hospitals etc.]
(ii) Firms [Private, Semi-government and government. These are better better clients in terms of paying capability].	Space & infrastructure and access to facilities. [Here the location criteria is not is not the same as above, as with increase in level of infrastructure, the rent increases. The rent commensurate with the level of facilities and the level of services.]
(iii) Institutions [Health, business etc. These have specific area requirements depending on their functions].	

3. LAND DEVELOPMENT SYSTEM

This is the system arising out of market forces. This also has some agents to perform certain activities, and thus forcing the land to develop. These are :

Agents	Activities
(i) Pre-plan Developers	Buying and selling land in the periphery of the existing towns in the direction in which the city grows.
(ii) Developers	Buying and selling land to get maximum profit. The developers can reach only 2 to 3% of the population as they have a high rate of selling. The developers will not built low cost houses or small shops. Land developed by the developers can only benefit a small percentage of population. So, this creates a spurt of illegal construction of houses and slums, as common people cannot afford to pay for the land developed by developers.
(iii) Consumers	Maximize the value of resources. There is a range between which a family can invest.
(iv) Financial Intermediaries or Brokers	To get maximum commission, they hike the rate of land. When housing is in scarce, the brokers profit, this is also linked with the land policy.
(v) Public Agencies.	

2.8 LANDUSE THEORIES:

There are three types of theories which describes the landuses in a city. These theories are:

(i) DESCRIPTIVE THEORIES : Describing the Urban situations, not necessarily the solutions.

(ii) EXPLORATORY THEORIES : Exploring certain aspects of Urban life, but not in totality.

(iii) SPECULATIVE THEORIES : Which try to bring the modern concept of planning by foreseeing the future. They are not tested.

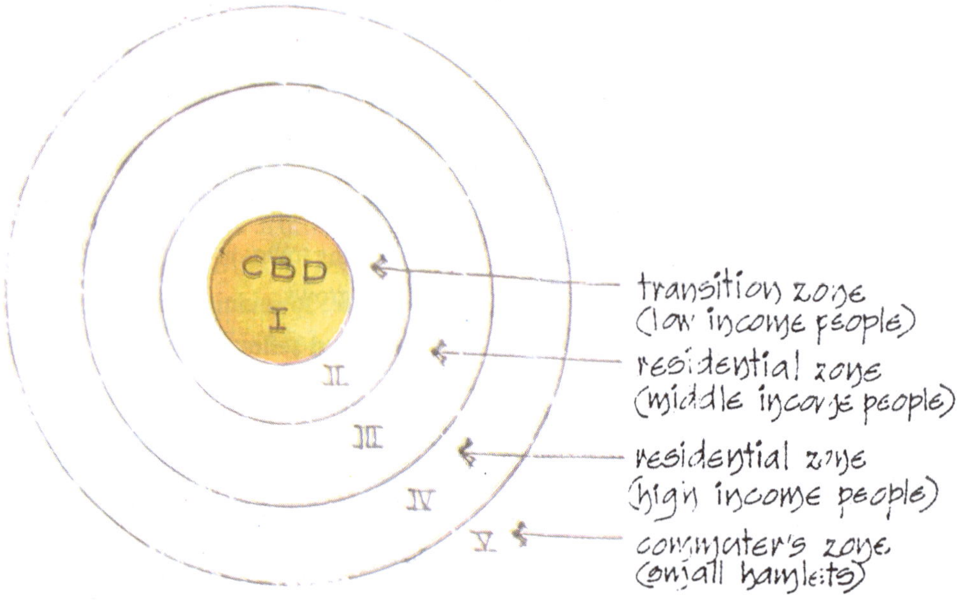

Fig. No. 19(a) : ***Parks & Burgess theory (1925)***

(i) DESCRIPTIVE THEORIES :

These are formed during 1925 to 1945. These theories were mostly developed in the U.K. and the U.S.A. These attempted to find out certain relationship between the city and its population. Here, the concept of transportation (of resources) developed a new meaning. Thus, developed the concept of inter-relationship and interaction through various actions.

(a) PARKS AND BURGESS THEORY (1925) :

This theory postulates that a city grows in concentric circles. The innermost core in the Central Business District or the C.B .D. followed by a transition zone of low income population. The third zone is comprised of independent residential area of middle income people and the fourth zone has residential areas of rich people. The fifth (outer) ring consists of Commuter's zone comprising of small hamlets who depend on the city for business.

This concept was based on a study done on 119 American cities. It imagined development of city as a monocentric phenomenon. It only describes the city suggesting no remedial measure, (vide Fig. No. 19a).

(b) SECTOR THEORY OF HOMER HOYT (1939) :

This theory is an extension of Parks and Burgess' Theory and it says that the zones may not be concentric but they may come out in wedges. These wedges are irregular in shape and size. This theory also fails to justify for multicentric cities, (vide Fig. No. 19b).

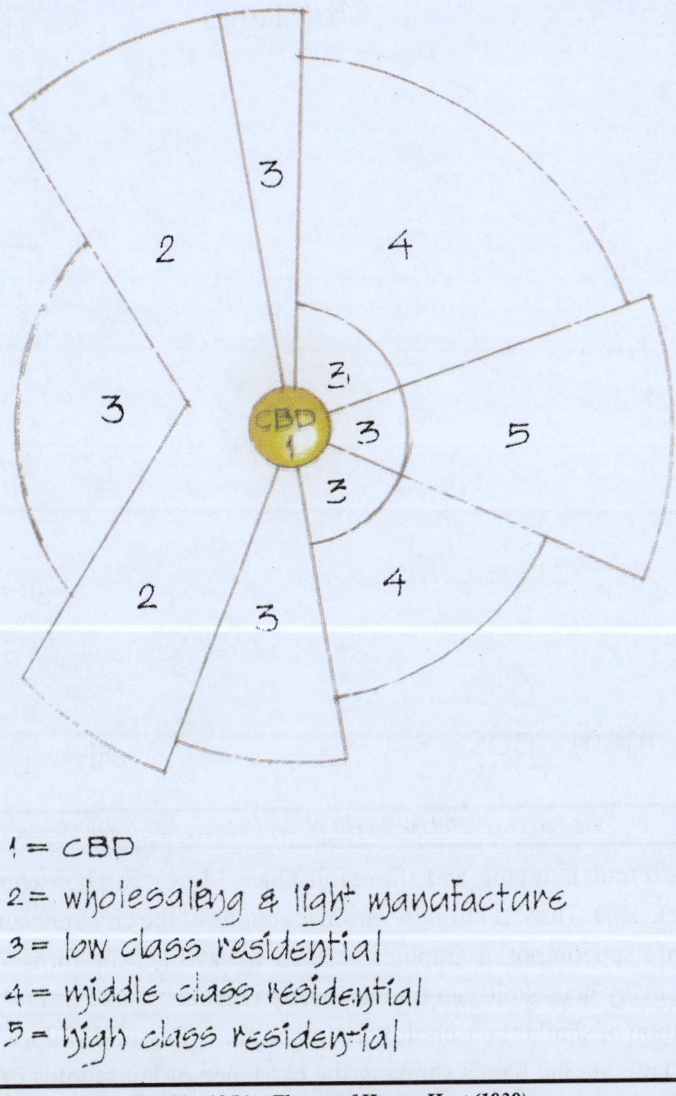

Fig. No. 19(b) : *Theory of Homer Hoyt (1939)*

(c) MULTIPLE NUCLEI THEORY OF CHAUNCY HARRIS AND ULLMAN (1945) :

This says that the cities may have multiple nuclei (or C.B.D.s) but the oldest C.B.D. may remain as the mother nucleus with higher moral and social values; but the newer centers may fetch higher rent because of higher infrastructural facilities. This theory works on the principle of 'gradation', (vide Fig. No. 19c)

(ii) EXPLORATORY THEORIES :

(a) CLASSICAL THEORY OF LANDUSE BY W. ALONSO (1964) :

According to this theory, a householdfinds its own location in a city depending on the rent structure As the rent structure very from the C.B.D. to the surrounding areas, a household fits

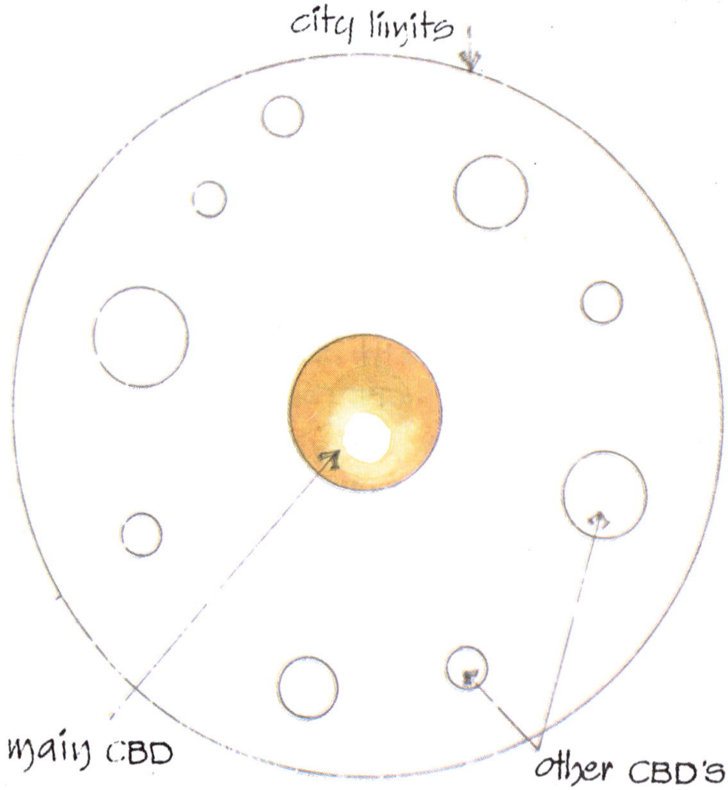

Fig. No. 19(c) : ***Multiple Nuclei Theory Chauncy Harris and Ullman***

itself only where it finds a suitable and affordable place. More rich a household is, more closer it will come to the C.B.D. Thus, if through various sections of the city 'rent curves' or 'bid curves' are taken, and are superimposed graphically, then we will get the form of solid, undulating hillock. On this, every household can find its location on the basis of its rent paying capacity. If the point of location of similar households are joined along the solid 'hillock' then, we will get a circular band. These are the bands showing the existence of households of various economic levels existing in the city .If a household improves his rent paying capacity by cutting down of some of its expenditures, then its rent paying capacity increases, and it comes closer to the C.B.D. (vide Fig. No.19d).

(b) A TRANSPORTATION ORIENTED THEORY OF LONDON WINGO JR. (1961)

Wingo said that a city is an accumulated demand for transportation. The trips made from a residential area for different purposes, are in a limited number compared to the trips made in a commercial center. Industrial areas and recreational areas have more trips but there is an inflated demand for these trips at a particular period of time. If we can assign certain values of these demands created by specific landuse and systematically analyze them, then we can put them together and form a systematic transportation in the city.

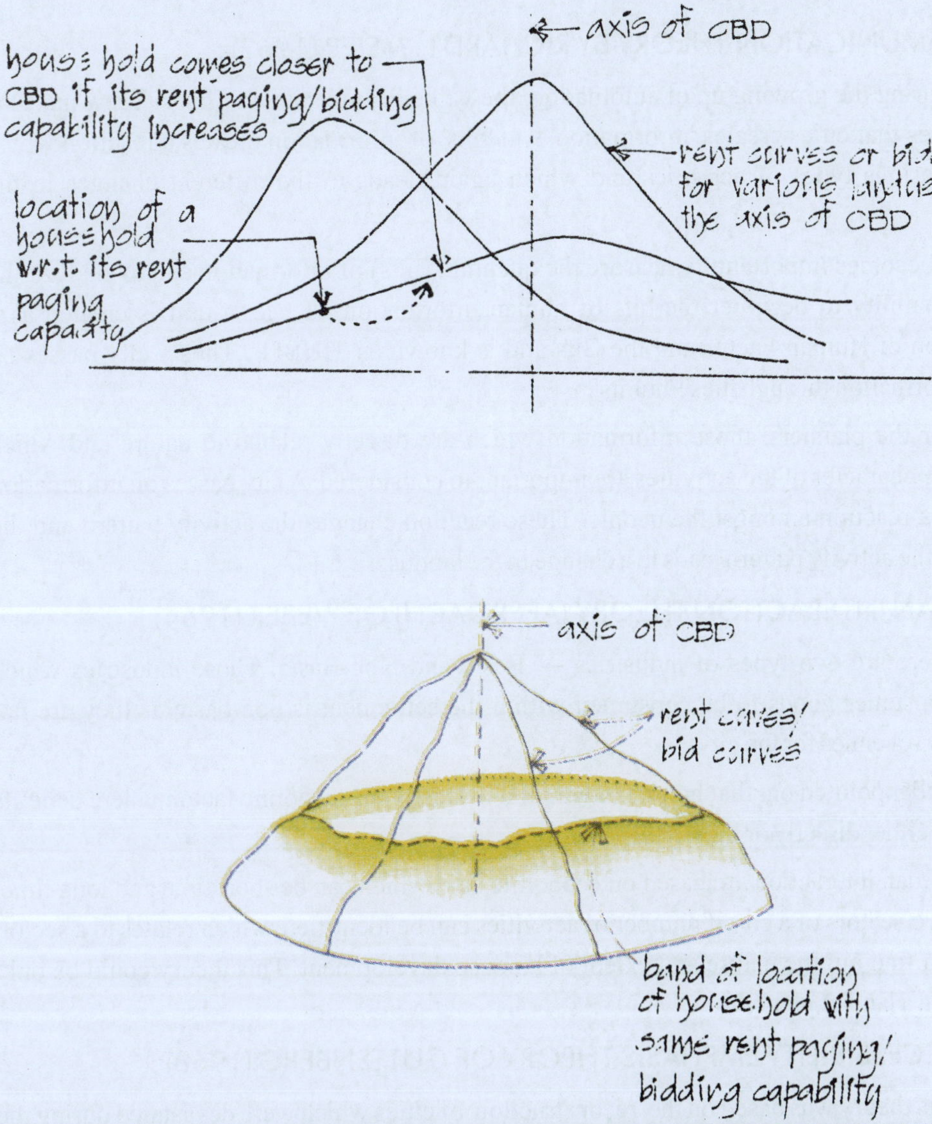

Fig. No. 19(d) : *Classical Theory of Landuse by W. Alonso*

(c) SOCIAL VALUES IN PLANNING THEORY BY WALTER FIREY (1947) :

This theory says that the pre-requisites like the services, houses etc. may and may not meet the demand of the human factors. When the systems are running in a perfect manner, there is hardly any room for personalized services; that is there is a state of 'anonymity'. But everybody wants a little more of his share — which is the basic human nature.

A city is for the human beings and the human beings are not for the city. Hence, there should be some amount of 'flexibility' for the people and we should not move towards anonymity. It is important to find the nature of this flexibility.

(iii) SPECULATIVE THEORIES :

(a) COMMUNICATION THEORY BY RICHARD L. MEIER (1962) :

During the growing up of automation, the value of Information Theory grew up. This theory states that cities creates 'information' and 'bits' of information creates a reaction which leads to various types of activities and which again, leads to the different changes in the landuses.

It becomes important to measure the quantum (bits) of information and to analyze the human capability to decipher the bits of human information. Hence, a unit is created as a combination of Human Factor and the Bits and is known as 'HUBIT'. Thus a city passes on several information through these 'hubits'.

For the planners, those information which are directly related to nature and which changes the character of the activities are important to considered. A city passes on information which has a reaction amongst the people. These reaction changes the activity pattern and the changes in the activity pattern leads to a change in the landuse.

(b) HUMAN INTERACTION THEORY (APPROACH) OF WEBER (1964) :

There are two types of industries — Basic and Non-Basic. Those industries which produce consumer goods to be consumed within the settlement is non-basic as they are not earning any revenues for the city.

Weber pointed out that human interaction is mostly on economic factor, where benefits are derived either directly or indirectly.

Human interaction are based on economic gain which can be short time and long time. Between two sectors of a city if number of activities can be identified, which relates to a sector, then we can find out the nature and extent of landuse development. This theory could be only conceived in a land of plenty.

(c) AN ACCESSIBILITY EMPHASIS THEORY OF GUTTENBERG (1968) :

This theory was based on the 'reconstruction' of cities which were devastated during the World wars. The main emphasis was the access to the facilities and services. The first job of reconstruction was to remove the mis-match. Guttenberg thought that establishing accessibility is the main point in Urban Planning.

Thus, the whole work was on establishing 'norms and standards'.

Guttenberg said that each landuse must be attributed to its accessibility. But the standards depend on the technological progress of the country and other social and economic factors. Hence, the standards developed in one country of a particular city cannot be borrowed for any other country or city; but they can be developed for a particular country and for a particular time.

2.9 SYSTEMS APPROACH TO PLANNING :

By 1942, the various postulates put forward by the planners complicated the situations of a city far beyond the idea by which planning can take place. The core of the city remained unexplored and the knowledge of the city was fairly ambiguous. The architects formulated a group of well designed buildings in the centers of the cities and related them with one another and the rest of the city was left to the common man to build their buildings anywhere. This tarnished the image of Urban Planning completely. It was then thought that those who has money can built a city and city building was becoming a part of civic art and not a part of civic design.

To avoid this, certain logico-mathematical expressions were sought to bring the vague and ambiguous sectoral thinkings into a concrete relationship. The factor of 'user's satisfaction' ('y') is thus depicted as :

$$y = ax_1 + bx_2 + cx_3 + \ldots + \ldots$$

where, ax_r bx_2, etc. are number of amenities.

This expression simply means that User's satisfaction is the sum total of the number of amenities available to them. This logic is suitable for the allotment of housing areas in a city.

To formulate such mathematical models, two things must be made very clear.

1. The problems must he clearly understood to provide certain dimensions so that the logic of the relationships is clear.
2. A suitable mathematical expression may be chosen to explain the problem.

Another expression, generally known as the Gravity Model may also be discussed here. In this model, which is similar to the expression of gravitational pull between two bodies, the attractive potential of an activity (say a shopping complex) to various sectors of a city, is tried to be found out. The expression runs like:-

$$A = K(\, mm'/d^2)$$

where, A is the attractive potential (say of a shopping complex in a sector of a city), K is a constant, m and m' are the mass (population) of two different sectors of a city and d is the distance between the sectors.

Each of the mathematical model is under the condition that every Other thing will remain the same. But in true life, things cannot remain constant. So, these models are of limited use as they are not dealing with variables but they are dealing with probabilities. This is the finite limitation in using the mathematical units in planning.

Thus, system approach to planning is adopted to do away with these problems.

An 'Approach' is that philosophy of planning process which dictates the process, which in turn, compliments the approach.

The following are the characteristics of system approach to planning.

1. A 'System' is set of elements and their attributes (attitudes and aspects). So, in System Approach Application, an understanding of the attributes of the elements is necessary.
2. System Approach organizes the planning operation into sequential steps.
3. Division of responsibility of individual projects is also done in system approach. Thus, a fragmentation of the task is done which brings in the concept of accountability.
4. Organizing the distribution of resources in such a manner that each task or project gets enough share of resources for its smooth implementation. Thus System Approach brings forth a proper strategy of resource distribution.
5. System Approach, at any given time can help in re-adjusting the goals, objectives and targets.
6. It also helps in organizing the post-implementation feedback by taking the :

   (a) User's Reaction

   (b) Evaluation of the strategies to find out whether they are proper or not.

   (c) Evaluating the applicability of the project by the community.

   The reasons for monitoring during implementation are also to be recorded in the approach. Planners, thus, must be able to define the trade-off. If no records for the reasons for the changes are kept, there would be 'gaps' between the intention at the time of formulation of the plan and the realities of the present day. These gaps are formed because of lapses or the unforeseen adjustments (like natural calamities etc.)

   The important schedule in the Approach is the time schedule. So, in planning process a 'cushion' period is kept. A tight schedule is prepared to begin with. This schedule, then, must be complimented to the 'cushion' period, to take care of unforeseen calamities.
7. A systematic record of each major and minor decisions which are taken during the currency of a project must be kept which helps in arriving at a documented case study; which in turn, helps in understanding the theoretical input as well as recalibrating some of the models that has been used in the Planning Process.
8. It sets out a metaprocedure for planning.

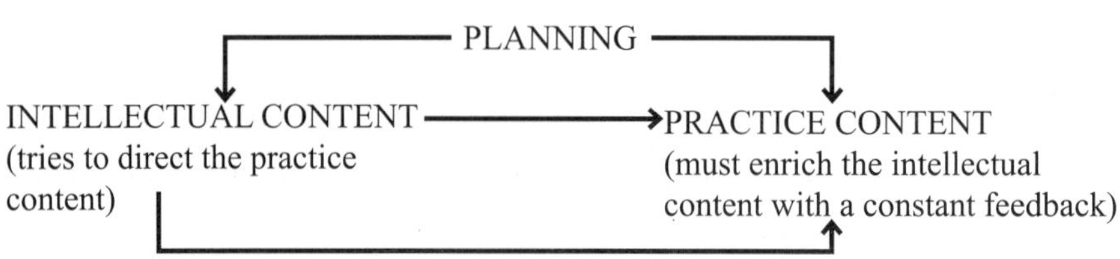

System Approach is applied on a planning process and it is not a planning process by itself. It acts as an envelope to cover up the planning process or we can also say that it comes as a discipline to orient the planning process.

2.10 MIXED SCANNING APPROACH OF A. ETZIONI (1968) :

This approach was suggested at the third approach of decision making (comprehensive approach and incremental approach being the other two.).

The whole approach is divided into four stages:-

1. STRATEGIC LEVEL

(i) To list down all possible alternatives to a problem situation including those which appear infeasible.

(ii) To examine briefly all the possible alternatives and reject those which are :-

(a) Infeasible due to lack of resources.

(b) Which appear to violate the basic values of the decision maker.

(c) Which appear to violate the basic values or interest of those whose support may be necessary for implementing the plan.

(iii) To examine in detail and repeating the process as described in (ii) above, and rejecting more alternatives.

(iv) To repeat the process described in (iii) above in a still greater detail any one solution (or alternative) is reached at.

Thus the Strategic Level is the process of selection by legitimate rejections.

2. BEFORE IMPLEMENTATION

(i) If possible the implementation is fragmented into several sequential steps (e.g. as an administrative rule).

(ii) If possible the commitment for implementation is divided into several serial steps (e.g. as a political rule).

(iii) When possible, the commitment of assets is divided into several serial steps and maintain a strategic result (e.g. as a utilitarian or management rule).

(iv) If possible, the implementation is arranged in such a way that costly and irreversible decisions appear later in the process than the less costly and the reversible ones; (as when implementation is started, new information may be coming and the costly decisions can be changed).

3. REVIEW WHILE IMPLEMENTING

(i) To scan at semi-encompassing level after the first subset of increments have been implemented.

This may not give any idea about how the system is functioning but we get an idea whether the work is compiled properly or not. If they 'work', then the scanning is continued at larger intervals; and Full Overall Reviews can be taken at proper time.

Thus, this approach works under a 'CONTROLLED SYSTEM' and the advantages of the Full Overall Reviews are that, we come to know about the system's weaknesses and whether each implements are co-related towards the end design. Here, a 'critical path' of each project is made and it is superimposed with overall dates. This is known as 'Mixed Programming Strategy'.

Thus, this approach does not allow one to discover the failures after it has taken place.

(ii) To scan in an overall manner when several set of implements have been implemented.

Thus, we asses the project schedule at a full running stage.

(iii) As per (i) and (ii) above, we can reset the project, its goals, objectives, targets etc.

4. FORMULATION OF RULES FOR ALLOCATION OF ASSETS AND TIME BETWEEN VARIOUS LEVELS OF SCANNING

The number of scans are fixed depending upon the size of the project; it is to be remembered that if the number of reviews are more the project may get delayed.

ADVANTAGES OF MIXED SCANNING APPROACH

(i) This metaprocedure can correct the planning process if it is going wrong. .

(ii) Mixed scanning approach concentrates on the management of plan implementation which has been lacking in other processes.

(iii) Mixed Scanning Approach provides ample scope for the use of 'Descriptive' as well as 'Predictive' models in the planning process.

(iv) This approach combines the gains of Rationalistic and Incremental Approach together.

DISADVANTAGES OF MIXED SCANNING APPROACH

(i) The deficiency of this approach is too much dependency on higher order technology.

(ii) This approach depends on a centralized co-ordination, while in our country, we have a decentralized management system at the sectoral level.

MIXED SCANNING AND URBAN DEVELOPMENT PLAN

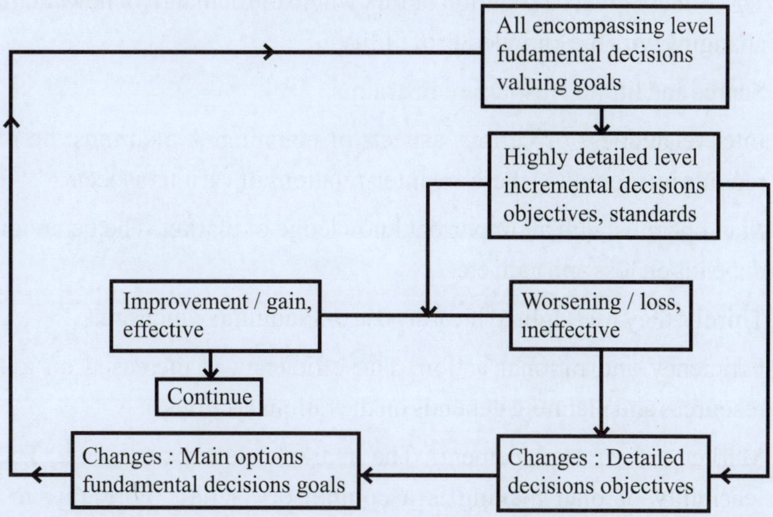

2.11 CHOICE THEORY OF PLANNING OF PAUL D AVI DOFF & T.A.REINER. (1962) :

In all the processes described before, there was an absence of a complete theory. Thus, to have a theory which is complete is all aspects, this theory was advocated by these two planners with economic background.

They defined 'Planning' as:' A process of determining appropriate future action through a sequence of choices'.

Therefore planning is meant for implementation and so, we must identify all the aspects which may create problem while implementation of the plan. Hence, the 'action' is the goal of this theory. At every stage of planning, there is a 'sequence of choices'. These choices must be complimented with 'information'. One must have a system of updating information, as they are available. Thus this theory ask for an information system which can be regularly updated. Establishment of Information System must precede plan.

Secondly, they gave the definition of 'environment' surrounding the planning; i.e., the environment within which planning takes place. This environment is characterized by :

(i) The individual Preferences. Individual preferences are taken together and they are being added and statistically reduced to any average pattern. But this pattern cannot give the general opinion of the people because information are got from persons having individual preferences. This is known as 'SUSPECT TREND ANALYSIS'.

(ii) Change or variation in preferences.

(iii) Each goods or services produced follow the dictum of 'DIMINISHING UTILITY'— these create a situation of flux where the demand for new facilities keep on changing with the knowledge of utility.

(iv) Scarce and limited resource constraints.

(v) Inter-relatedness of various aspects of planning. A plan must be implemented module by module as there are inter-relations of various aspects.

(vi) Man operates with an imperfect knowledge of market. The behavior of markets depends on loss and gain etc.

Thirdly, they highlighted the Purpose of Planning, which are:

(a) Efficiency and rational action. The efficiency is increased on availability of resources and planning depends on the rational actions.

(b) Market aid or replacement. The market must be studied. The economic feasibility of plan making is a compulsory action. Formally, in the pseudo socialistic environment, private bodies were not entrusted in our country; very recently, we have started thinking of'privatization'.

(c) Change or widening choice. People's choices change over a period of time. The only purpose of planning is to bring in certain change in the environment, building conditions etc. Development is defined as the sum total of growth and change. Growth is physical but change may be psychological.

Fourth, they enumerated the Planning Characteristics; which also serves as the Indicators for selection of a proposal. These characteristics are :

(a) The achievement of'ends'.

Planning process has two parts, one is the 'ends' or objectives and the other is the 'means' by which planning is done. The 'means' must be rational by which it is most likely to achieve the goals. In the process of defining 'means', it may be found that the 'ends', are unclear, hence the 'ends' must be clear and achievable within the foreseeable future.

(b) The exercise of choice.

Choice must be taken in each stage of making the plan.

(c) The orientation to the future.

The plans must be flexible to accommodate changes in the economy. There should be scope of relative expansion of potentials in each of the landuses.

(d) The planning actions.

The actions are sometimes postponed and then they create problems. Hence, the actions must be planned to aid the forces of development in proper time.

(e) Comprehensiveness.

Comprehensiveness is in contradiction with the other four characteristics as the other four demands quick action and it is obvious that comprehensiveness tends to slow down a system. Anyway, comprehensiveness is also necessary to formulate the long time goals of planning. So, Davidoff and Reiner suggested that all the plans must have a comprehensive perspective.

The advocators of this theory warned the planners against the facts and values. Many a time planners act like a social reformer, completely changing the ways of social life. Facts about the life of the people should be got and it should be thought that whether the life-styles should be changed or not. The social values may not result out of facts — it may also be based on some advertisement campaign.

In 1962, when this theory was first published, there was a criticism by John Dakin, and then the authors gave a rejoinder. The criticisms put forward by Dakin were:

(a) It neglects the spatial character of planning. John Friedman said that 'there cannot be any theory of planning which is divorced from city building'; it means that a city cannot be designed Without esthetics.

(b) It draws upon the social imbalances in the society without addressing itself to the root causes; i.e historicity, economic distribution, exploitation by the markets, racial discrimination etc.

(c) Human factor has been given a backseat. There are three things which are fixed for a society. The political boundary, the defined amount of resources, time, money etc. and the physical realities of men and environment, their customs, norms and lifestyles. Within these three things a planner makes a number of alternatives. But a planner cannot break any of these physical realities. The alternatives generated by the planners is such that many choices are removed by the first two girdles. Therefore, the choices cannot be a true social choice, and these also may not be the most appropriate choice. Planners in a communist country cannot follow the choice theory as there, everything is given (dictated) by the State and no choice is entertained.

(d) Validity of choices may also be questioned. The ability to choose depends upon our background and the planners having different background can opt for different vehicles for progress.

(e) It does not accommodate change in the political and economic situations.

2.11.1 Application of Choice Theory in Indian Context :

This theory can be applied in Indian context if the following points are clarified and understood. These points are:

(i) Relation between planning and politics.

This requires a clear strategy and this must be a part of the planning process. By too much of political interference, no planning model or methodology can bring the application of the Choice Theory in practice.

(ii) Planning Education/ Consciousness.

Common man in India is not planning conscious. It seems that Plans are made to take away the rights from the landowners. Thus the promotion aspect of planning is completely overshadowed by the controlling aspects of it. Thus education of planning must be incorporated within the masses.

(iii) Planner's Role.

The planners are suffocated with the political and resource canvases. There must be a co-relation between the Official Agency Plan and the Private and People's Plan. The Plans must be written in understandable language.

2.12 ADVOCACY AND PLURALISM IN PLANNING :

PLURALISM : It is in the context of the society. A pluralist society is where there are different sectors. The dividers being religion, economic status, social ideologies and the way of life. The economic status of two persons may be the same but the expenditure patterns may be totally different. Pluralism is the character of modern society and every modern society is in a state of flux. Modern society differs from older society by the level of tolerance. Thus anything and everything becomes intolerable and thus changeable within a very short period of time.

So, the most important role is of one who can make different social groups begin to see a common good and that man is the **'advocate'.** [Of course not to be confused with the lawyers]. The role of the 'advocates' is advocacy. Modernization does not mean removal of different voices and ideologies but to harmonize different ideologies for a common good'. The role of advocacy is increasing day by day.

The two basic concepts of society, like advocacy and pluralism is brought into a uniform code of conduct which is, 'Planning Consciousness'. This is the first step of Advocacy and Pluralism which is commonly known as **'Public Participation'.**

A society defines the **'Goals'**, and the planners suggest the **'means'** to achieve the goals and the society, after sometime, experience the **'result'.**

2.12.1 Public Participation :

Thus, planners must prepare the plans and educate the society about the plan and obtain the help and co-operation from the society.

In Advocacy planning, the people must be involved and the involvement must be institutionalized.

In 1969, a committee was made at New Delhi with the chairmanship of A.M. Skeffington. The recommendations made by the committee were:

1. People should be kept informed about the plan and how their areas are going to be affected. It was suggested that every residential area can have a 'Block' and a number of such 'Blocks' can create a 'Local Planning Committee'. These committees can be from local level to area level and the issued brought up the planners can be 'structured' in these committees.
2. Involve local people at the data collection stage. In this way, certain unforeseen areas will be emerged out which may support the conceptualization of the plan or formation of standards.
3. Representation should be analyzed and the respondents must be informed whether their suggestions being accepted or not.
4. Local Planning Authorities should explain alternative investment opportunities to the people. There should be more understanding of the Planning which will help in promotion.
5. Mass education regarding the benefits of planning must be given. (This was practiced in Milton Keynes, England, and was found successful).

The process has also got some drawbacks such as:

(a) This process sometimes is regarded as a process of 'objection' rather than a process of 'suggestion'.

(b) Perfect democracy is not available as there is always certain degrees of limitations imposed by the government.

(c) It is time consuming process.

(d) The process may get distorted by vested interests.

3 THE MASTER PLANS AND THE LANDUSES

In India, 'Master Plans are made for the cities of Madhya Pradesh and for Delhi. For Uttar Pradesh there are Development Plans, for Madras there is Structure Plan and for Calcutta there is Basic Development Plan. All these are different form one another in their context.Here, we will be only discussing about the nature of the Master Plans and the landuses that are provided in them.

Master Plans are prepared by the analysis of social and economical data applied to whole of the city. Plans and estimates are prepared, approved by proper authorities and the programs of works are made.All these form the parts of a Master Plan. These Plans should be elastic and non-rigid in nature to accommodate any necessary change in course of time.

3.1 PRINCIPLES FOR PREPARING THE MASTER PLANS :

(a) It should provide for all major objectives for the development of the city and still be economical.

(b) Provisions for changes during the implementation of the Plan should be kept.

(c) It should be sound, practical and inspire civic interests, devotion and loyalty.

(d) No over crowding should be there with assurance of health care, recreational facilities and amenities.

(e) Proper traffic management systems should be incorporated in it.

3.8 REQUIREMENTS TO IMPLEMENT THE MASTER PLANS :

(i) Co-operations with Local officers.

(ii) Public support.

(iii) Help from Central Government.

(iv) Proper legislation—Police force.

3.3 POLITICAL POSITION OF MASTER PLANS :

MINISTER (for Urban Development) → TECHNICAL ARM → DEPARTMENT OF TOWN PLANNING (The secretary of which is an IAS Officer, forming a part of the ministry) →OFFICE OF THE CHIEF TOWN PLANNER.

THE MASTER PLANS ARE LEGAL DOCUMENTS. The financing ofthe Master Plans are done as per the State and the Central Five Year Plans.

3.4 ORIGIN OF THE MASTER PLANS :

1919 : First Development of the Master Plans.

1940 : (During 2nd: World War), Plans for roads, water supply and sanitation were made.

1946 : The Health and Survey Development Committee (Bhore Committee) was set to advice the British Government for the remedy of the three deficiencies (i.e. the roads, water supply and sanitation) of the Indian towns. During this period many coastal towns and Nagpur were developed.

1947 : An Act to make the Master Plans were passed. The schemes made under this Act were known as **Urban Improvement Schemes.** The first **Planning Organizations** were the **Urban Improvement Trusts,** which were formed by a number papers as written by Patrick Geddes for the towns of Southern India. These Trusts failed as there were no co-ordination between one scheme and another.

After 1947, in all British colonies, Master Plans were made mandator.

Upto 1955, only Rehabilitation towns were made like.Nilokheri, Faridabad, Rajpura etc.

Only in 1952, Chandigarh was planned.

From early 1940s, a number of towns were planned in eastern India. Dr.Koenigsberger planned towns like Jamshedpur, TELCO, TISCO plans, Bhubaneswar etc.

Before 1955 Town Planning was under the Ministry of Health, when the Central, Regional.and Urban Planning Organizations were created. This was done to help the State Governments to create the Town Planning Departments. Then a number of towns came up like that of Nangal, Bhilai, Rourkela, Durgapur etc. To date (1990), we have 118 New Towns completed.

3.5 REASONS FOR FAILURE OF MASTER PLANS :

(i) The preparation of Master Plans took 3 to 3'/2 years and by that time the data collected becomes obsolete.

(ii) Adoption of faulty techniques (like that of statistical projections formulae for the future needs etc.)

(iii) The 20 years perspectives that were taken in the master plans proved to be too much for the proper implementation of the Plans.

(iv) Most of the Master Plans were made very general, which needed detailed Zonal Plans.

(v) No serious planning was ever done.

(vi) The plan making body has no say in the implementation of the plans.

(vii) No qualified Town Planners were available at the time when most of our Master Plans were done.

(viii) The Master Plans evoked a negative feeling for the Government Sector and the individuals.

By 1960, something new was felt to be made. Then **Calcutta Basic Development Plan** was made. It did not have any landuse plan, which was a departure from the Master Plans. The entire Calcutta Metropolitan District has 33 municipalities, 3 corporations and 1 Development Authority. So, different programs were handed over to different agencies, which created more confusion.

3.6 STRUCTURE PLANS:

In 1971, Structure Plans were developed in Britain. It consisted of 5 parts:

(a) Structure Plan: Which is a Policy Plan and the boundary of which coincides with the administrative boundary.

(b) Local Plans : The parts of the Structure Plans which are to be developed.

(c) Subject Plans: For an area with special emphasis as declared in the Structure Plan.

(d) Regulation 8 Plan: For specific areas with typical problems, which can go against the provisions of the Structure Plan.

(e) Programs : It is a written document which deals with a subject matter which are necessary for each of the Local Plans.

In 1975, Madras Metropolitan Development Authority prepared the first Structure Plan in India; and in 1985, the concept National Capital Region (NCR) was started.

Advantages of Structure Plans :

(i) In the Master Plans there is no justification for the option adopted but in the Structure Plans the Planner is forced to document the options.

(ii) The Structure Plans has the ability to deal with diverse planning situations.

(iii) It allows the Physical Plans to be sensitive to the policy contents of the National and State Level Plans.

3.7 PLANNED LANDUSE TYPES AND THEIR HIERARCHY :

There are seven basic types of landuses as mentioned before. These basic landuses can be further divided into various subgroups in hierarchical orders depending upon the type of the town and the need for the hierarchical orders. This hierarchy as formulated by the Delhi Master Plan (1962) is listed hereunder :

RESIDENTIAL		
RD	Residential area with density upto	: 250 pph
		: 300-500 pph
		: 550-750 pph
		: 800-1000 pph.
RF	Foreign Missions	
COMMERCIAL		
C_1	Retail and General business	
	Central Business District/Sub-Center District.	
	District Center	
	Community Center	
	Non-hierarchy Commercial Center.	
C_2	Wholesale and warehousing.	
	Warehousing and Depots.	
C_3	Hotels.	
MANUFACTURING		
M_1	Light and service industry.	
	Flatted Group Industry.	
	Service Center.	
M_2	Extensive Industry.	
RECREATIONAL		
P_1	Regional Park	
P_2	District Park	
	Major Children's Parks	
	City Forest	
P_3	Playgrounds, stadiums, and sports complex.	
	Divisional Sports Center.	
P_4	Historical Monuments	
TRANSPORTATION		
T_1	Airport.	
T_2	Rail terminal, Integrated Passenger Terminal,	
	Metropolitan Passenger Terminal.	
T_3	Rail Circulation	
T_4	Interstate Bus Terminals, Bus depots.	
T_5	Truck Terminal; ntegrated Freight Complex; Parking	

T_6	Road Circulation	: 90m R/W
		: 75m R/W
		: 60-64m R/W
		: 45m R/W
		: 30-36m R/W
		: L.T.R. Route.

UTILITY

U_1	Water (Treatment Plant etc.)
U_2	Sewerage (Treatment Plant etc.)
U_3	Electricity (Powerhouse.sub-station etc.)
U_4	Solid Waste (Sanitary landfill etc.)
U_5	Drain.

GOVERNMENT

G_1	President Estate and Parliament House
G_2	Government Offices
G_3	Govt. Land (Use undetermined).

PUBLIC/SEMI-PUBLIC

PS_1	Hospital
PS_2	Education and research University; University Center; College
PS_3	Social and Cultural;Cultural Complexes; Socio-Cultural Institutions.
PS_4	Police Headquarters and Police Lines
PS_5	Fire Stations and Headquarters
PS_6	Communication; Transmission Site, Transmission Center; Facility Center (PS 1 TO PS6)
PS_7	Cremation and Burial Ground
PS_8	Religious

AGRICULTURAL AND WATER BODY

A_1	Nursery
A_2	Green Belt
A_3	Rural
A_4	River and Water Body.

4 TRANSPORT PLANNING

4.1 INTRODUCTION:

The landuse of transportation is one of the most important of all the other landuses because it not only links one landuse with the other but it also is the most expensive of them all. It provides access to a space which is a pre-requisite of the usefulness of any space. Without the ability to enter, leave or move within it, and to receive and transmit information or goods, space is of no value, however vast in size, or however rich in resources it might be.

A City is primarily a communication net of roads, paths, rails, pipes,wires etc..which is related to the activity pattern (i.e.landuse), which again is dependent on the economic and cultural life of the people. Hence the communication networks vary from one city to another and the overall development of a city is decided by it.

The use intensity of any circulation system depends mainly on the density, F.A.R., and the employment sector of the users.

The circulation system is mainly concerned with the FLOWS that takes place through them. The flows may be of different types , like that of people, goods, wastes, services and information which takes place through some CHANNELS and by some MODE.

The Channels are mostly consisting of NETWORKS and TERMINALS & INTER-CHANGES.

4.2 NETWORK CHARACTERISTICS:

There are two basic characteristics of the networks namely : the FORM and the CLASSIFICATION & HIERARCHY.

4.2.1 The Form :

It is the FORM of the network that gives the SHAPE of any city. There are various patterns of roads a few of which are enumerated here under with examples. (vide Fig. No. 20).

(i) Grid-Iron :

Roads crossing at right angles. Example: Chandigarh, India.

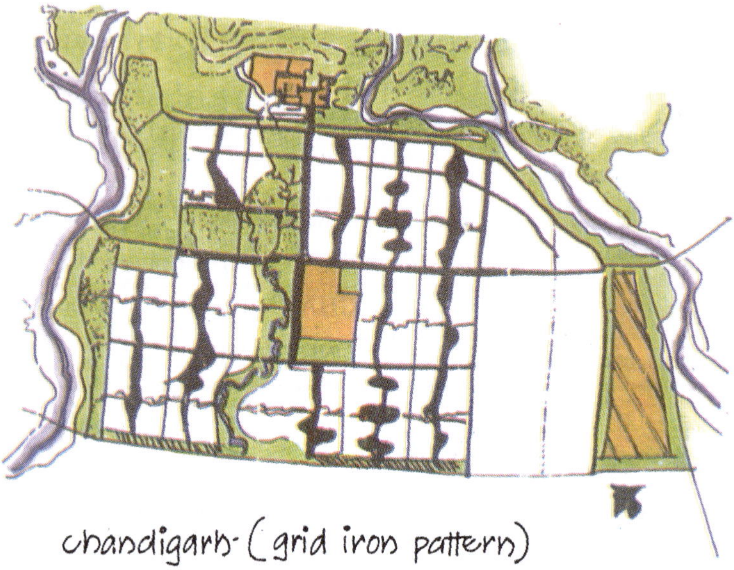

Fig. No. 20(a) : *Form of Networks — Roads and City Shapes*

Advantages:

(a) Regular Plot sizes are got.

(b) Easy flow of traffic is maintained.

Disadvantages:

(a) Monotony of drivers as all crossings look similar.

(b) Close placement of roads creates too many intersections.

Fig. No. 20(b) : ***Form of Networks — Roads and City Shapes***

(ii) Regular and Diagonal:

Grid pattern with diagonal crosses. Example : Washington D.C.; U.S.A.

Advantages:

(a) Monotony is broken.

(b) Landmarks and Nodes are created.

Disadvantages : Plot sizes becomes irregular.

(iii) Radial:

Roads spreading out from one point. Example:Karl Sruhe, France.

Advantages:

(a) City gets a prominent center.

(b) It gives a direction growth of the city.

Disadvantages:

(a) Trapezoidal plots are got.

(b) Through traffic is increased.

(c) Connections of the radials are felt necessary as the distance from one road to other goes on increasing as one travels on any road.

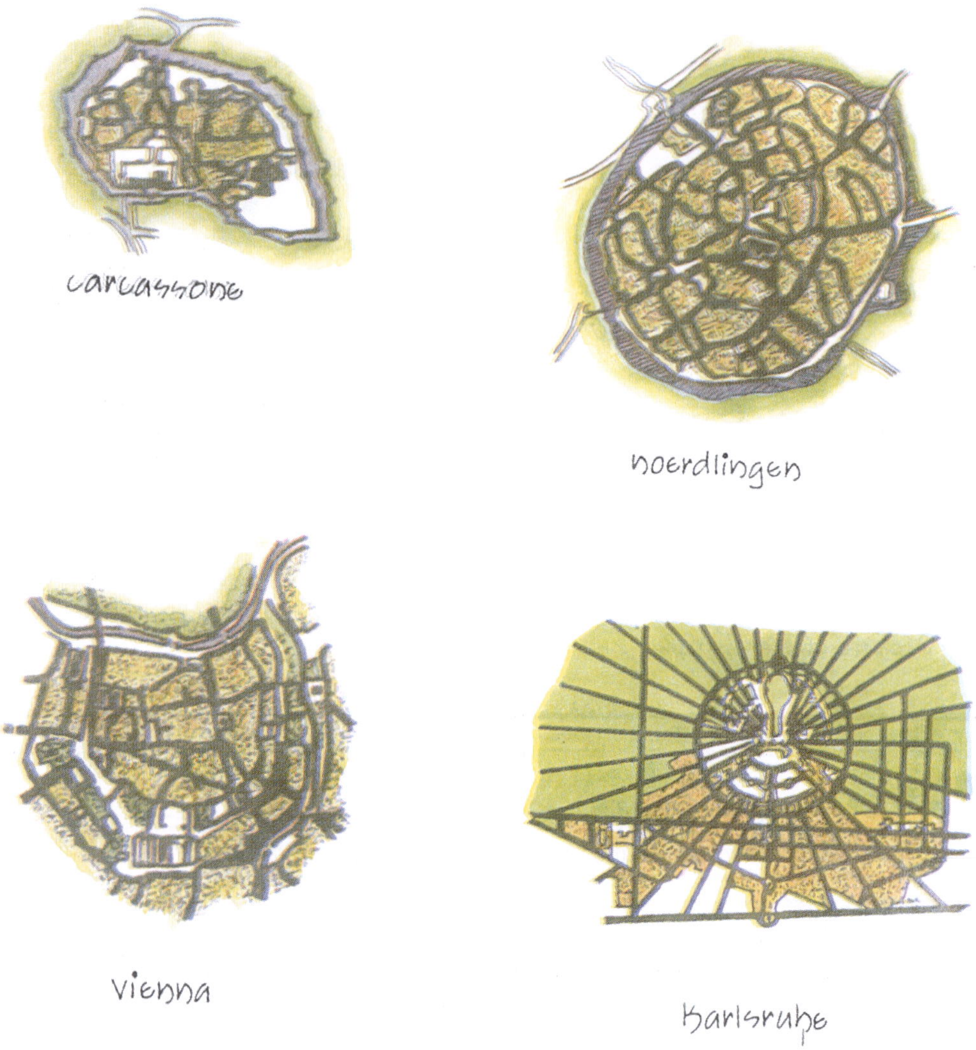

Fig. No. 20(c) : ***Form of Networks — Roads and City Shapes***

(iv) Radio-Centric :

Radial roads with concentric circles. Example : Vienna, Austria.

Advantages :

(a) Easy connection between radials.

(b) Other advantages of Radial pattern are also got.

Disadvantages :

(a) Too many concentric rings increases intersections.

(b) Trapezoidal plot sizes are got.

(v) Radial and Rectangular :

Combination of radial and grid-iron pattern. Example : New Delhi, India.

Advantages :

(a) Prominent nodes and landmarks are got.

(b) Regular plot shapes are got.

Disadvantages : Similar to both radial and rectangular patterns.

(vi) Organic: Branching type of streets developed as per needs. Example: Carcassone, France.

Advantages :

(a) Provides a new vista at every step.

(b) Meets the public demand in the best possible way.

Disadvantages :

(a) Difficult for automobiles.

(b) Irregular plot sizes are got.

(vii) Other Types :

A. Combination of regular and irregular : Example : Edinburgh; England.

(In many of the examples, the regular patterns were overlapped over irregular types).

B. Irregular Medieval: Example : Varanasi,India; Nashik, India.

These roads provided security from the invading enemies in the medieval period.

C. Linear : Example : Calcutta, India; Mumbai, India.

4.2.2 Classification and Hierarchy :

Classification of roads can be made in 3 ways viz. Grain, Specialisation and Hierarchy. Here we will discuss the Hierarchical classifications only.

The major classification as per hierarchy is of 2 types :

(i) INTER-CITY Roads : The roads connecting in between 2 cities.There are various categories of Inter City roads,like:

(a) Expressway : Having high density uninterrupted traffic with speed 100-120 Kmph. No parking is allowed on these roads and provisions for surveillance facilities are there. These roads are separated by 5 Kms. The ROW is 60m. Example : Ahmedabad-Vadodara Road (80Km. long).

(b) National Highway : ROW :50-60m. Roadway : 9-12m.

(c) State Highway : ROW:50-60m. Roadway : 9-12m.

(d) District Roads : ROW:25-30m.Roadway : 9m.

(e) Village Roads : ROW : 10m. Roadway : 7.5m.

Note : Village roads are also included in Inter-city Roads as these roads are not found within a city.

(ii) INTRA-CITY Roads : The roads which are within the city limits.The various categories of Intra City roads are:

a) Arterial Road : Usually a continuous route for through traffic. The design speed of these types of roads is 180 Kmph. The ROW is 50-60m.The arterial roads are separated by a distance of 1.5 to 5 Kms.

b) Sub-arterial: These are continuous route for through traffic but offering a lower level of traffic mobility than the arterial roads.

The design speed is 60 Kmph. ROW is 30-40 m.

(p) Collector street: These roads are meant for collecting and distributing traffic from and to local streets and also for providing access to arterial streets. Design speed is 50 Kmph.ROW is 20-30m.

(d) Local streets : A street primarily for access to residences, business or other abutting property. Design speed is 30 Kmph. and ROW is 10-20m.

4.3 DEFINITION OF TERMS:

RIGHT OF WAY (ROW): The horizontal distance from one plot line to another plot line in front of it.

CARRIAGEWAY (C/W): The metaled portion of ROW to take the vehicles.

VOLUME OF TRAFFIC (V): Total traffic flow at a point.

CAPACITY OF ROADS (C): Number of vehicles that pass through a highway in one hour under ideal conditions.

OPERATIONAL SPEED:The speed at which traffic flows on the road.The Operational Speed depends on the volume to capacity ratio of that particular road.

Note : In the Table (4.3.1) given below, as the V/C ratio increases,the operational speed on the road decreases.

Table 4.3.1

V/C	Operational Speed
0.1	Higher
0.5	Medium
0.8	Stop & Go
1.0	Unstable

LEVELS OF SERVICE (L.O.S.): It is the qualitative expression describing the traffic condition at a given state of roads.This depends on:i) Driver's Freedom

(ii) Operational Speed.

(iii) Volume/Capacity Ratio.

Table 4.3.2

L.O.S.	V/C	Condition	O. Speed
A	0.1 -0.3	Free flow	> 80 Kmph
B	0.3	Freedom affected	80 Kmph %
C*	0.3 - 0.5	Freedom restricted	60-80 Kmph
D	0.5-0.7	Stream behavior	40-60 Kmph
E	0.7	Stop & Go	20-40 Kmph
F	1.0	Forced flow	0-20 Kmph

Note: L.O.S. of type C is ideal for urban roads as the drivers are cautious to avoid any accident.

PASSENGER CAR UNIT (P.C.U.) : Traffic in the urban areas is always of mixed nature.The width requirement of the roads should be assessed on the basis of equivalent Passenger Car Units using the tentative equivalency factor as shown in Table 4.3.3.

Table 4.3.3

Vehicular Type	Equivalency Factor
1. Passenger car.tempo, auto-rickshaw, jeep, van or agricultural tractor	**1.0**
2. Trucks.bus or agricultural tractor-trailer	**3.0**
3. Motorcycle.scooter and cycle	**0.5**
4. Cycle rickshaw	**1.5**
5. Horse-drawn Vehicle	**4.0**
6. Bullock Cart	**8.0***
7. Hand Cart	**6.0**

Note: For small bullock carts a value of 6.0 will be appropriate.The equivalency factors given above are applicable only to through sections of urban roads in-between junctions.

4.4 INTERSECTION :

4.4.1 Design Criteria :

The Intersections are the places where the vehicles can change their directions. There are various criteria for designing the intersections. They are :

(i) Type of road
(ii) Priority of movement
(iii) Volume of Traffic
(iv) Allocation of land
(v) Existing shape of intersection (i.e. T-junction, Cross junction etc.).

4.4.2 Types of Intersections :

Intersections may be Grade Separated or Controlled (manually or mechanically). Grade separated Intersections can be of the following types: (Vide Fig. No. 21)

(i) Partially Grade Separated.
(ii) Fully Grade Separated.

Grade Separated Intersections may be again of various types namely:

*Simple Grade Separated.
*Universal Interchange.
*Braided or T-Interchange.
*Four level Interchange.
*Simple Interchange of freeway or highway.
*Clover-leaf interchange.
*Bel-Geddes Interchange.

Improvements of existing intersections could be done by channelisation of traffic with the help of 'Medians' and/or division islands.

The intersections which are controlled can again be of various types namely :

*Priority Intersections : Where one needs to stop, look and go. There is no control as such in these intersections.

*Manually Controlled: Where a traffic police is installed to watch the volume of traffic on the roads and allows the traffic of any one direction to flow.

*Signal Controlled: The cycle period of the traffic signals can be so arranged that too many vehicles need not wait for long for the 'Green' to come. The time period can be arranged on the volume of traffic on particular direction of the intersection.

*Area Controlled System (Link System): Where the signals along a particular stretch of road are so linked with one another that any vehicle moving at a specified speed along the road will not be stopped by 'Red' signal at any of the intersections along the road.

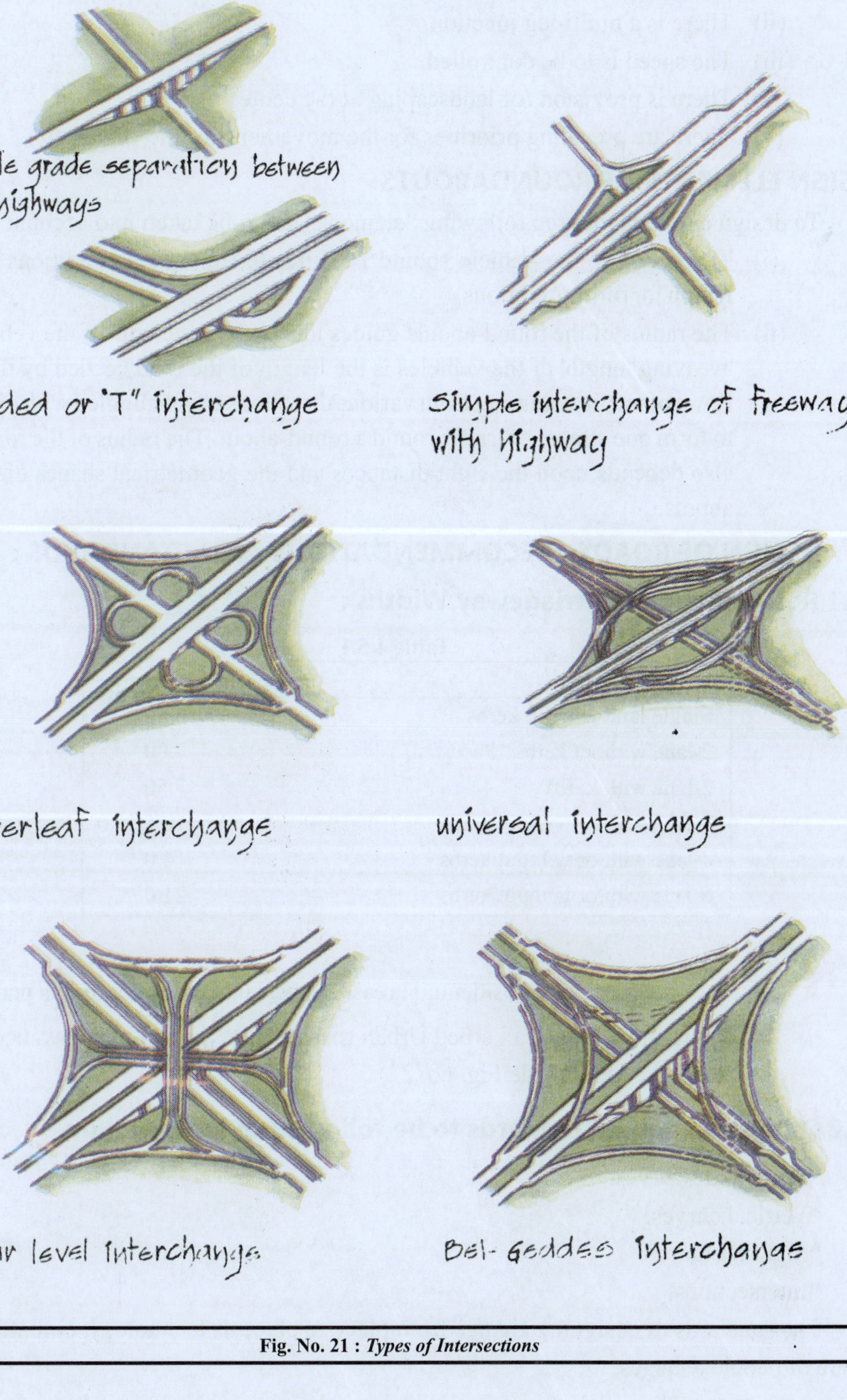

Fig. No. 21 : ***Types of Intersections***

*Space Sharing (Round about):They are efficiently used in places where :

(i) Free flow is necessary.

(ii) There is a multi-leg junction.

(iii) The speed is to be controlled.

(iv) There is provision for landscaping at the center.

(v) There are presiding priorities for the movements of the vehicles.

DESIGN ELEMENTS OF ROUNDABOUTS

To design a roundabout the following "elements' are to be taken into account:

(i) The speed of the vehicle should be 30Kmph for urban conditions and40-50 Kmph for rural conditions.

(ii) The radius of the round-abouts guides the 'weaving length' of the vehicles. The 'weaving length' of the vehicles is the length of the road needed by the various flows of traffic comming from various directions to get mingled with one another to form one stream of traffic round a round-about. The radius of the round-about also depends upon the sight distances and the geometrical shapes of the round abouts.

4.5 DESIGN OF ROADS : RECOMMENDATIONS AND STANDARDS :

4.5.1.Recommended Carriageway Widths :

Table 4.5.1

Description	Width (in m)
Single lane without kerbs	3.50
2-lane without kerbs	7.00
2-lane with kerbs	7.50
3-lane with or without kerbs	10.5/11.0
4-lane with or without kerbs	14.0
6-lane with or without kerbs	21.0

Notes :

1. For access roads to residential areas, a lower lane width of 3m. is permissible.
2. Minimum width of a kerbed Urban road is 5.5m. including allowance for a stalled vehicle. (Vide Fig. no. 22)

4.5.2 Other Geometric Standards to be followed In road design:

*Horizontal Curves.

*Vertical curves.

*Sight Distances.

*Intersections.

The standards of designing change from place to place as technology, climate and culture of people changes.

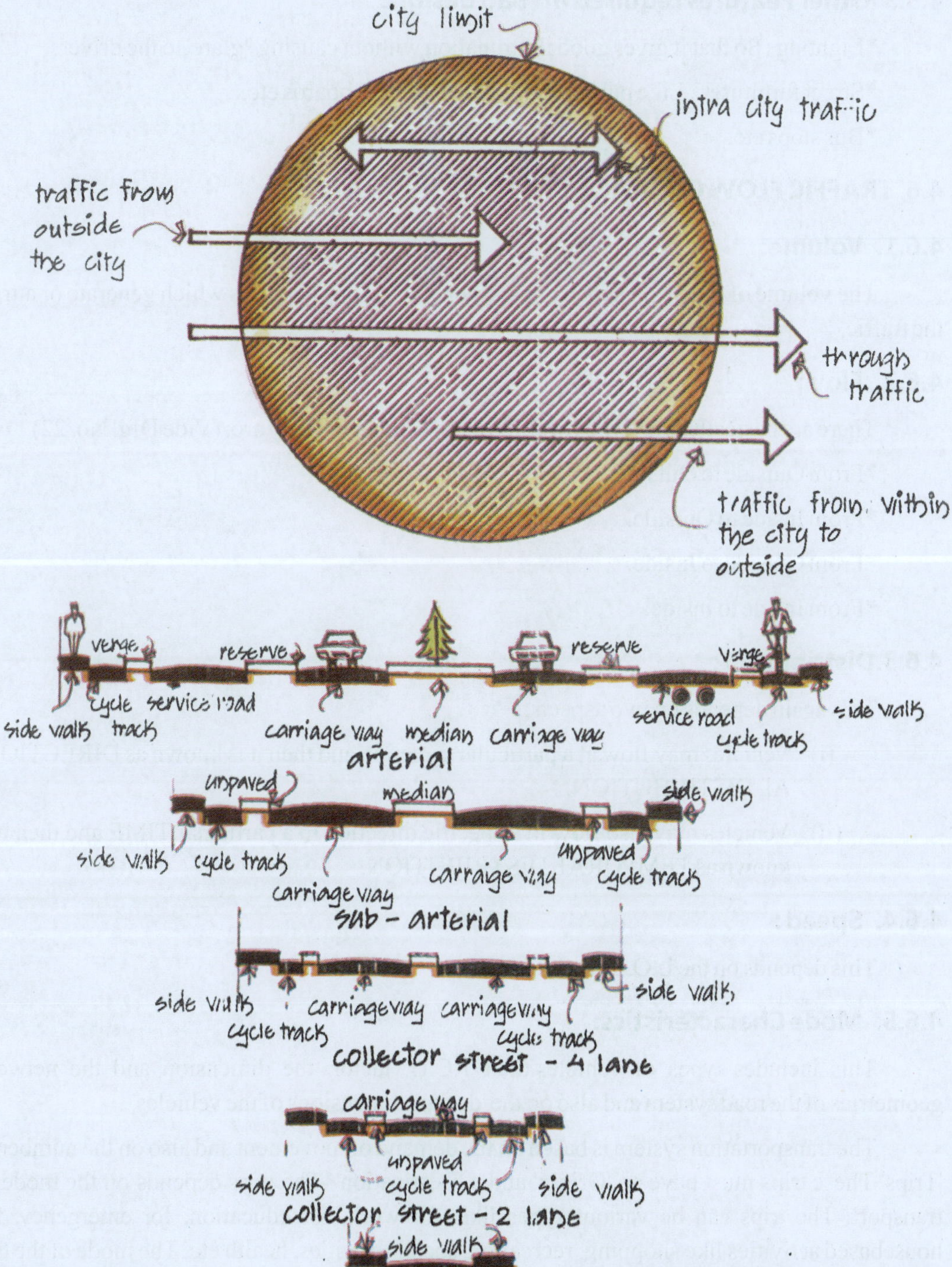

Fig. No. 22 : *Types of Flow of Traffic & Road Sections*

4.5.3 Other Features required in road design :

*Lighting : So that it gives good illumination without causing "glare' to the drivers.

*Street furnitures : Like public seats, advertisement boards etc.

*Bus stops etc.

4.6 TRAFFIC FLOW CHARACTERISTICS :

4.6.1. Volume:

The volume of traffic on a road depends on the various landuses which generate or attract the traffic.

4.6.2. Flow:

There are basically, four types of flow of traffic in a city, they are: (Vide Dfg. No. 22)

*From Outside to Outside.

*From Inside to Outside.

*From Outside to Inside.

*From Inside to Inside.

4.6.3.Distribution :

This again depends on two aspects;

(i) Vehicles may flow at a particular direction and then it is known as DIRECTIONAL DISTRIBUTION.

(ii) Vehicles may also flow in a specific direction in a particular TIME and then it is known as TEMPORAL DISTRIBUTION.

4.6.4. Speed :

This depends on the L.O.S; V/C Ratio etc.

4.6.5. Mode Characteristics:

This includes types of vehicles.their P.C.U. factor, the dimension and the network geometries of the road system and also on the 'design' dimensions of the vehicles.

The transportation system is based on the demand of movement and also on the number of 'Trips' These trips must have an 'Origin' and a 'Destination' .The trips depends on the mode of transport. The trips can be various types like, for work, for education, for emergency, for housebased activities like shopping, recreation, social activities, health etc. The mode of the trip can be met either by Public Transport, (like buses, rail, trams, etc.); Intermediate Public Transport (like auto-rickshaw, taxi etc.; and slow vehicles like rickshaw etc.); or by Other Than Public Transport (like Personified Slow Vehicles e.g. cycles and Personified Fast Vehicles like cars etc.)

4.6.5.1. Modal Split:

The percentage of the total demand met by Public Transport is known as 'Modal Split' ,i.e. if 40% of the total demands of trips is met by the Public Transport System of the city, then the Modal Split of the city is said to be 40%.

4.7. TRANSPORT SURVEY:

Transport survey may be done for assessing the number of vehicles that pass through a stretch of road which enables the planner to 'design' the roads and also to designate the location of the landuses as the 'trips' are generated by the 'landuses' only. Apart from this, surveying is also conducted to asses the area required for parking (the demand of which again depend on the landuses), particularly in the developed countries, where the personalized vehicle ownership is much more and thus parking of vehicles do create problems in the city centers and near the commercial areas, where land price is high on one hand and a lot of parking space is required at the same location on the other. Thus, Transport Survey may be broadly be divided into two types namely, Traffic Survey and Parking Survey.

Traffic Survey consists of 3 parts namely :

(i) Observation,

(ii) Assessment

(iii) Designing and Proposals.

Observation is mainly done of the congestion of the streets and a probable solution sought by taking into consideration the cost factor.

Assessment of the existing capacities of the roads can be done in relation to

(a) Carriageway Widths and

(b) Intersection Capacities. Assessment is also to be done on the 'KIND' of traffic that is flowing through the town. Hence, traffic survey is basically of 3 types namely :

(i) SPEED SURVEY : To check the speed at which the vehicles move through a particular stretch of road; for assessing the L.O.S .required and provide the same if found safe and economical.

(ii) VOLUME SURVEY : To asses the KIND of traffic that is flowing through the road and to find out whether the existing carriageway width is sufficient for the present traffic and how many more P.C.U.s it can sustain in future.

Here, a group of surveyors take a note of the kind of traffic that is flowing through a particular stretch of road on an ideal day.over a month or a fortnight, convert them to P.C.U.s by multiplying them with their respective equivalency factor, and an average of it is done to give a fairly good idea of the KIND of traffic that is flowing on the road.

(iii) ORIGIN-DESTINATION (O-D) SURVEY : To know how many of the incoming and outgoing vehicles have their Origins and Destinations in the town and how many are through traffic, just passing by the towns thus increasing the congestion on the roads.

O-D Survey can be done by a cordon of surveyors stationed at various points of intersections and the survey can be conducted by:

(a) Direct Interview, e.g.stopping the vehicles and asking them about their Origin and Destination.

(b) By observation of the registration numbers of the vehicles when they come in and go out of the town.

(c) By attaching tags to the vehicles when they come in he town and collecting the tags when they go out of it.

In all the-above mentioned three types of surveys, 'TIME' factor is also to be taken into consideration.If a vehicle take more time to pass through a stretch of road, it will mean that the vehicle has done some business on the way, and it was not just 'passing through' the city.

It is always not possible to conduct a 100% survey and 'Sampling' can be done which not only saves time but is also economical. In the developed countries home interviews are also done for traffic survey mostly to asses the number of 'Trips' a family takes and also information are collected about the 'Mode' of those trips.

Parking Surveys

By parking surveys, the planners want to know :

(i) The number of vehicles parked in the central area at different times of a day.

(ii) The number of frustrated parkers who wished to park in the central area but have parked somewhere else;

(iii) How many people habitually walk along the central area as they know that there is no parking space there;

(iv) How many, because of this reason, go to other central area of the town.

The results of the traffic surveys can be shown as :

(i) FLOW DIAGRAM showing the routes traversed by traffic in a town, and the amount of traffic in a given road at a given time.

(ii) DESIRE LINE DIAGRAM (which is independent^ the actual road system). It shows the main points of Origin and Destination of traffic and the relative quantity of traffic passing from each point of Origin and Destination.This gives a clear picture of the traffic pattern needed, (e.g. the traffic attracting and the traffic generating landuses and highlights the difficulties and opportunities of the actual road system got from the Flow Diagram.

5 HOUSING

5.1 DEFINITIONS:

The World Health Organization (W.H.O.) defines Housing as:

The concept of housing is an enclosed environment in which man finds protection and feels safe and secured from hostile forces and can function with increased comfort and satisfaction as regards privacy to the individual and his family. The environment must include all necessity, services, facilities, equipments and devices needed for physical and social well being of the family of the individual.

The basic inputs of housing thus becomes :

(i) Land, (ii) Materials and (iii) Finance.

Housing starts from one room and the combination of rooms creates a house.The science of arrangement of these houses with the provisions of all the necessary (physical and social) infrastructures is'Housing'.Thus the definition of a ROOM is to be understood very clearly to understand the subject 'Housing'.

A 'room' may be an enclosure having four walls and a roof

having a minimum dimensions of 2m. or 2.5m. and 2.4m. In height.

It may be of any material and it also may not have an opening other than the door .The National Building Code (N.B.C.) defines a room as:

A 'Room' must have an area of 7.5sqm. and a height of 2.6m. Openings must not be less than l/6th. of the floor area in hot-humid climate and l/10th. of the floor area in hot-dry climate.Openings(other than the door) have to be in the external area. lt must be made of permanent material.

Definition of Household:

Household is a number of persons staying together and taking meals from the same kitchen unless exigencies of work prevent them from doing so. There are three basic types of Households:

(i) SINGLE FAMILY HOUSEHOLD (consisting of social or blood relations),

(ii) MULTIFAMILY HOUSEHOLD (consisting paying guests) and

(iii) INSTITUTIONAL FAMILY HOUSEHOLD (consisting of hostels).

Occupancy rate :

It is the number of persons ;living in a room. Generally if the number exceeds 3, the room is said to be'crowded'.

5.2 OBJECTIVES OF PLANNING HOUSING AREAS :

The main objectives of planning Housing areas are listed hereunder:

(i) To provide sufficient living space within and around the houses to meet the need of families of different sizes and compositions.

(ii) To arrange the houses and to control the spaces in between them in such a way as to:

(a) Secure optimum conditions of environment in relation to climate, human comfort and health.

(b) To Provide adequate space for private and communal enjoyment of outdoor living.

(c) To secure healthy sanitary conditions.

(d) To provide adequate and safe access.

(e) To provide visually pleasant and stimulating surroundings.

(iii) To ensure that the people who live in these houses will be provided with all 'Community Facilities' conveniently located to meet their daily needs; e.g.market, shops, schools, recreation areas, open spaces etc.

(iv) To locate the housing conveniently in relation to urban areas as a whole, providing ready access to town and district centers, principal areas of employment, recreational areas and other housing areas.

5.2.1Elements of Housing :

There are various 'Elements' that are to be considered before designing an area for housing. A detailed study of these elements and their implications on the buildings and the implications of the buildings on these elements are extremely necessary before we go for planning an area for housing. These elements are:

(i) Access of the site.

(ii) Orientation of the site. (i.e. the north line, from which we can get the Solar Path and thus calculate the solar radiation on the structures).

(iii) Direction of the wind flow.

(iv) The soil condition.

(v) The contours of the site.

(vi) The natural drainage of water.

(vi) The flora (existing trees) on the site.

(vii) The noise level at the site.

(viii) The set-backs and the other building bye-laws.

(ix) The existence of any building (which may be important from heritage point of view)

(x) The level of smoke and other pollutants in the site.

(xi) The view and vista available from the site, (vide Fig. No. 23.)

5.3 TYPES OF HOUSING :

The various types of housing are listed hereunder:

(i) **DETACHED DWELLING HOUSE**: The advantages and disadvantages are as follows:

(a) Climatologically it is the most suitable as we can place the doors, windows and other opening as we like.

(b) Incremental development is possible.

(c) It is costlier.

(d) More side open space is required resulting in lesser number of houses along a stretch of street.

(ii) **SEMI-DETACHED HOUSES** : There is a common wall between two houses. Though economic, climatologically it is not as good as detached houses.The major disadvantage is that the two buildings are to be developed simultaneously as the cost of common wall is borne by two persons.

(iii) **COURTYARD/PATIO TYPE** : This type of houses are common in all hot dry climatic areas.These are introvert in planning. It ensures the maximum plot coverage. Generally there are no common walls in between the houses.

(iv) **ROW HOUSES** : Traditionally this type of housing is utilized for the economically weaker section of the community.If set in proper manner considering the wind direction, solar radiation etc. then row housing can be very comfortable to the dwellers.

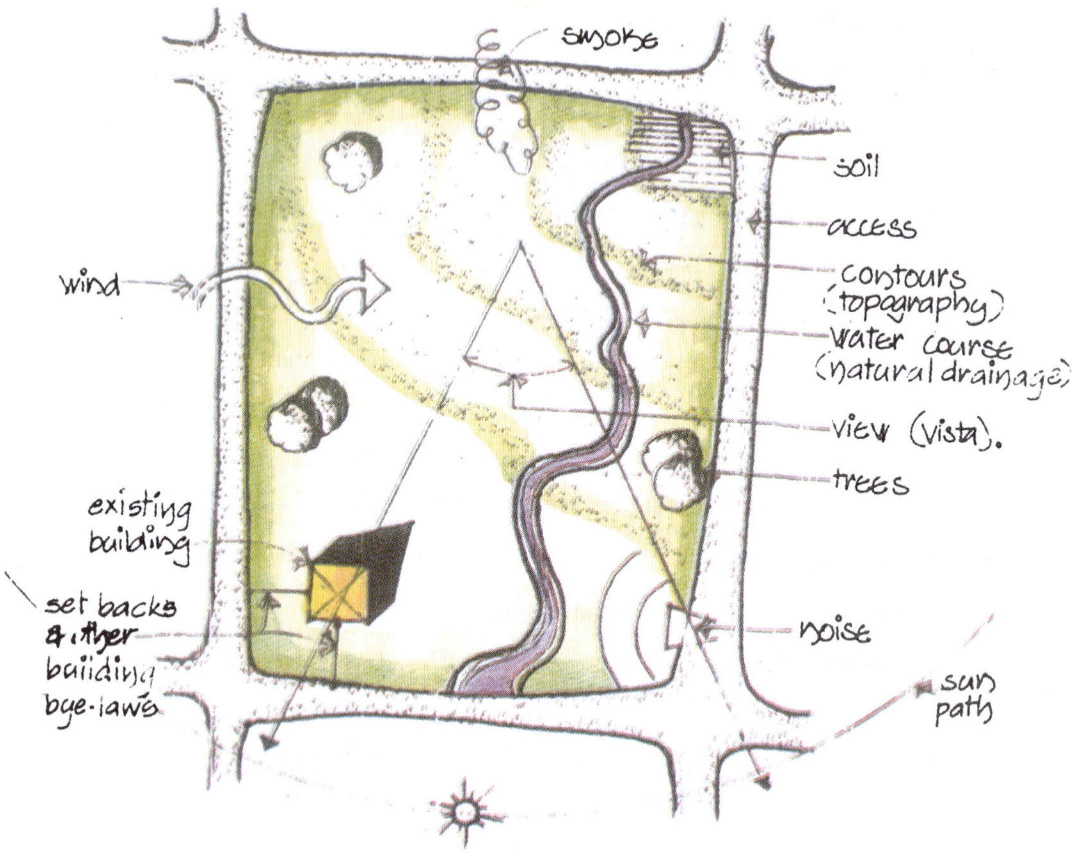

Fig. No. 23 : *Elements of Housing*

(v) CORE HOUSES : This is one of the latest of the housing trends. Here, a 'nucleus' or 'core' with the basic four walls and a roof is made ,where a family can stay .The important criteria is that it can grow in time. It is also called 'EXPANDABLE DWELLING'.

(vi) APARTMENT HOUSES : It is a ready-made house where there is access to 3 or more houses by a common lobby .Climatologically it is most suitable if placed in proper orientation. No change, alterations or improvements could be made in these type of houses.MULTISTORIED APARTMENTS are common where F.A.R. is higher.Here, fire proofing and fire fighting requirements must be met adequately.

From the economical point of view, types of housing are categorized for different income levels.The table given below (no.5.3.1) illustrates the various types of housing for the various income levels as decided by the Government of India, (vide Fig. No. 24)

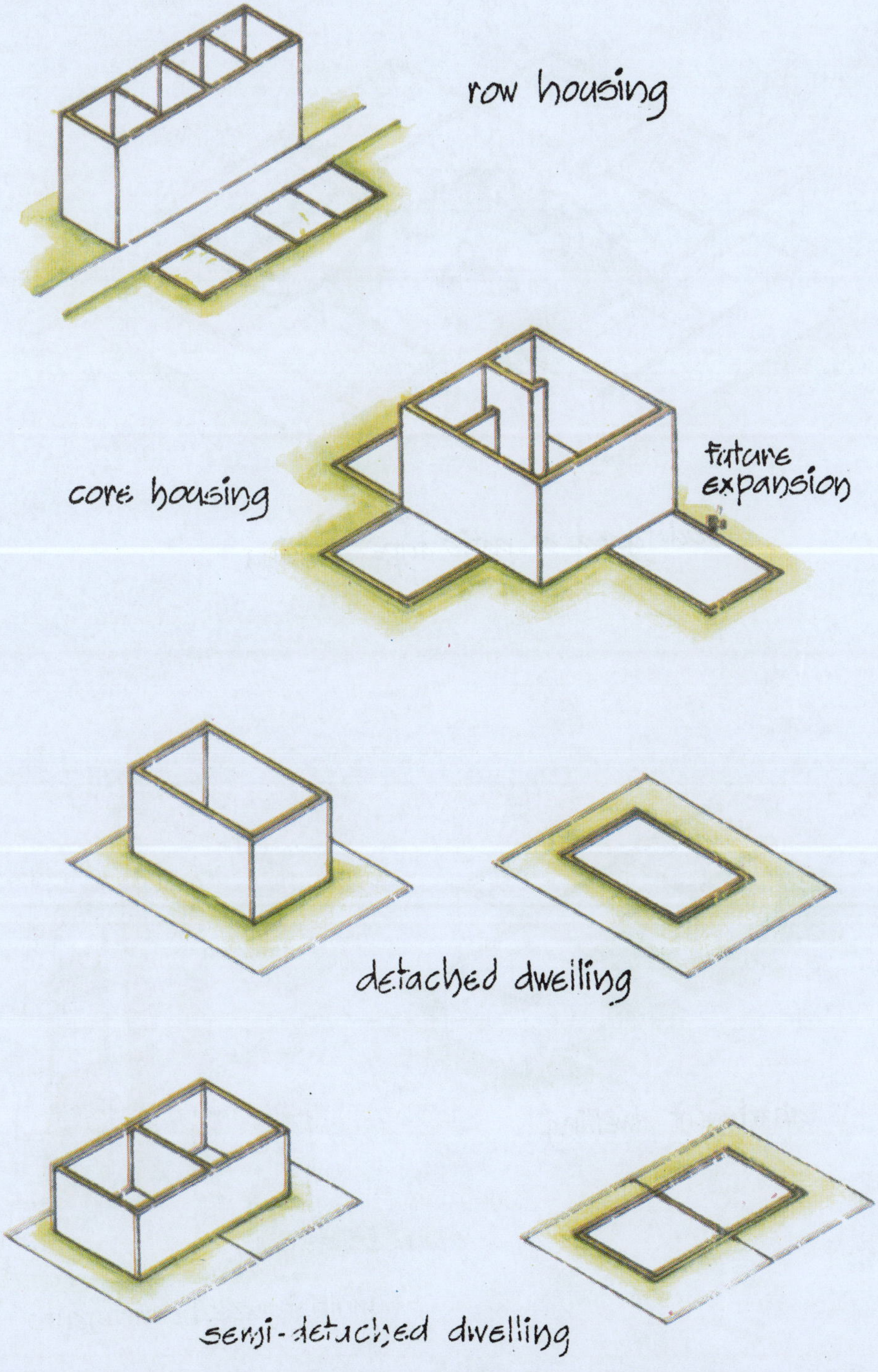

Fig. No. 24(a) : *Types of Housing*

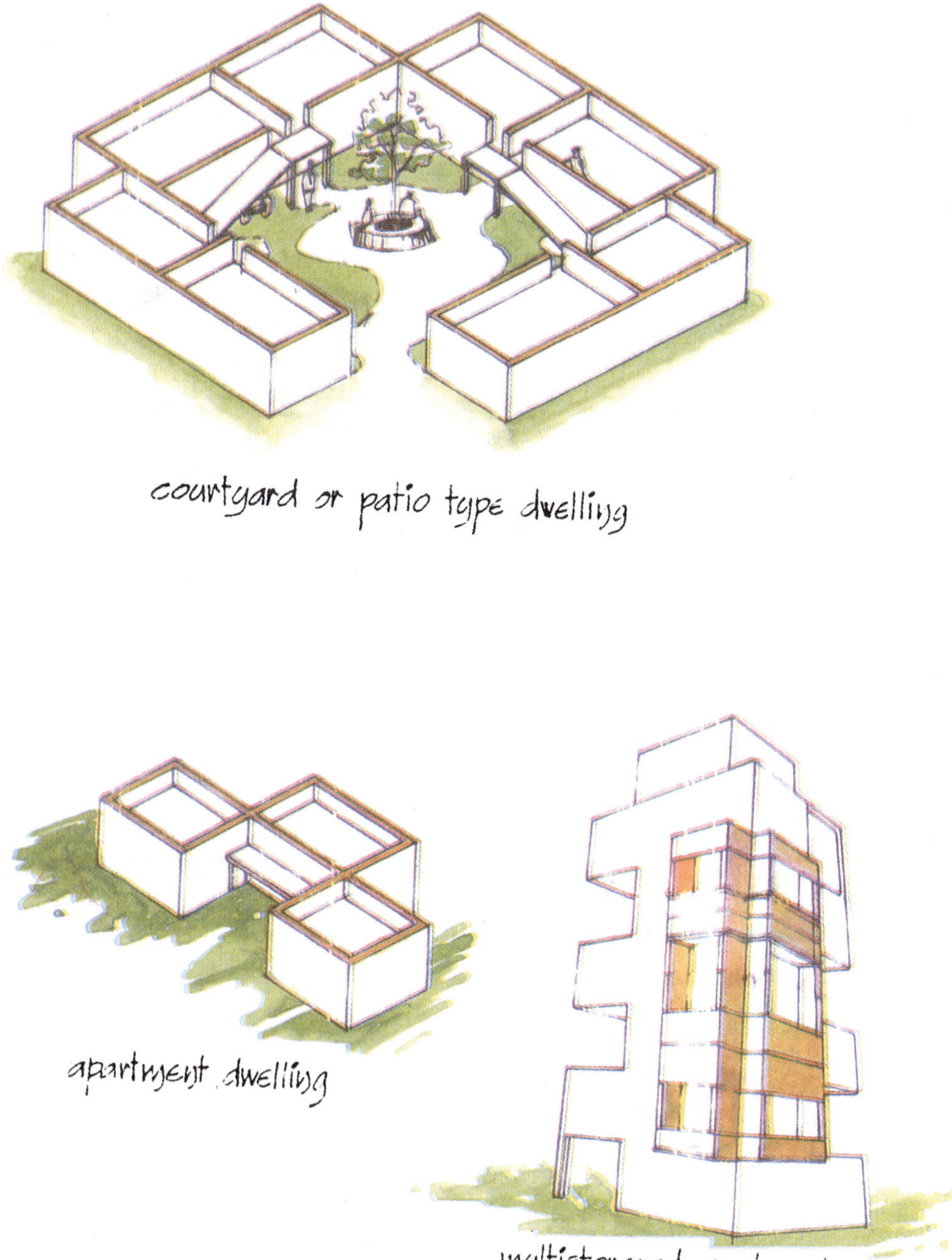

Fig. No. 24(b) : ***Types of Housing***

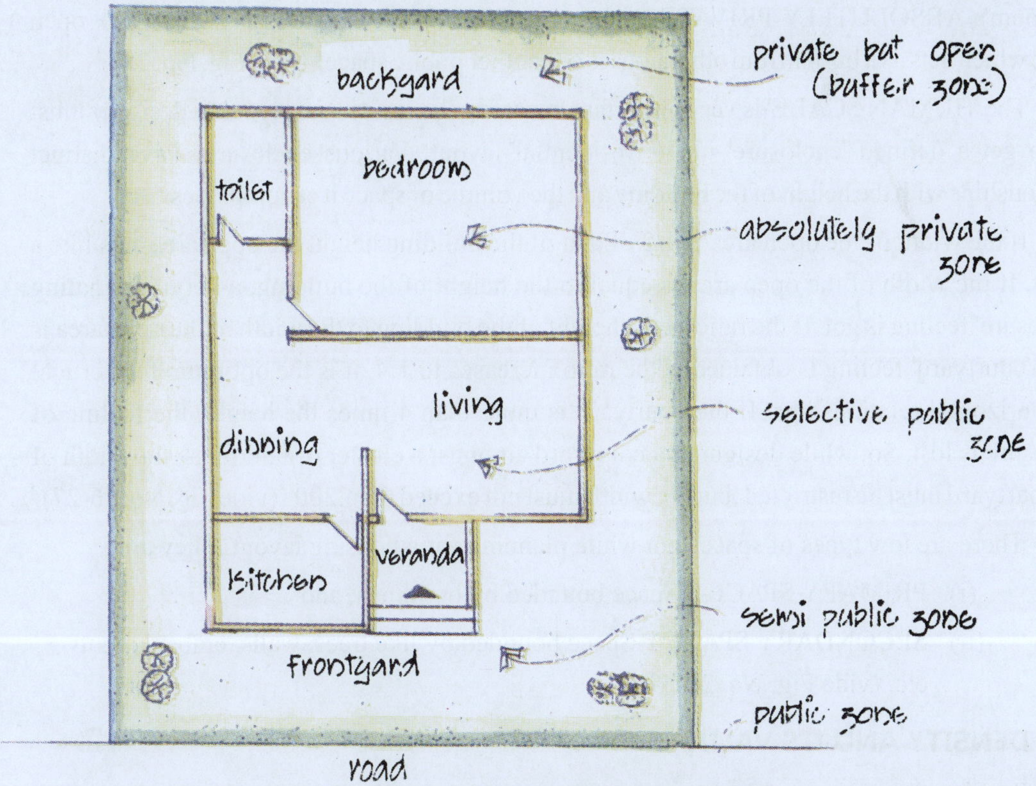

Fig. No. 25 : *Zones in a Residential Building*

Table 5.3.1

Types of Housing	Income Level per year (in Rs.)
EWS (Economically Weaker Section)	< 25200.00
LIG (Low Income Group)	25201.00-54000.00
MIG (Middle Income Group)	54001.00-87600.00
HIG (High Income Group)	87601.00 +

Note : No plot should be less than 30 sqm. in area.

5.4 HOUSING LAYOUT:

Elements to be considered for Housing Layout:

These are already described in Art. no. 5.2.1 and Fig. no. 23 on pages 110 & 112.

Zones in layout of Residential Units:

A residential unit has five zones from the access road to the back yard. These zones are: PUBLIC(Road); SEMI-PUBLIC(Front open space or garden); SELECTIVE PRIVATE (Liv ing room); ABSOLUTELY PRIVATE(Bed room); PRIVATE BUT OPEN AREA(Back open space which acts as a buffer from other houses or another public space).(vide Fig. No. 25)

The 'HUMAN SCALE' is very important in zoning the residential layout. A designer must try to get a defined 'enclosure' in the residential layout. Various enclosures have distinct relationships with the height of the building and the volume of space it encompasses.

If the width of the open area is 1/2 of that of the building height the open area acts like a street. If the width of the open area is equal to the height of the building, a very dominating 'enclosure' feeling is got. If the ratio of the height of the building to the width of the open area is 1:2, a'courtyard' feeling is obtained.If the ratio increases to 1:4, it is the optimum for vertical and horizontal relationships. If the courtyard is more than 4 times the height, the feeling of enclosure is lost. So, while designing a courtyard amongst a cluster of buildings, the width of the courtyard must be restricted ;but the width must not exceed 61m.200'. (vide Fig. No. 26-27)

There are tow types of spaces got while planning for a housing layout. They are :

(i) PRIMARY SPACE—Space bounded by buildings; and

(ii) SECONDARY SPACE—Space bounded by like trees, walls, embankments etc. (vide Fig. No. 28-29).

5.5 DENSITY AND ITS VALUE :

Density is a measure of the intensity of the occupation of land. Various types of densities are in use like Population Density, Dwelling Unit Density, Gross Residential Area Density, Net Residential Area density etc.

Value of Residential Density :

(i) To provide measures of rate of landuse, that could be used to estimate the total land requirements for the purpose of the community.

(ii) To provide an instrument for assessing the level of utility and infrastructure, thereby maintaining the living standards within the housing area.

Criteria for selection of the density ievel :

In order to determine the most appropriate housing density in a particular area it is necessary to examine, assess and reconciliate a wide range of local factors. They are:

(i) Health Factors:Which is taken care by water supply, sanitation and waste disposal, availability of sunlight,air, ventilation, internal living spaces etc.

(ii) Social Factors:Like Private Open Space, privacy, protection, community facili-ties etc.

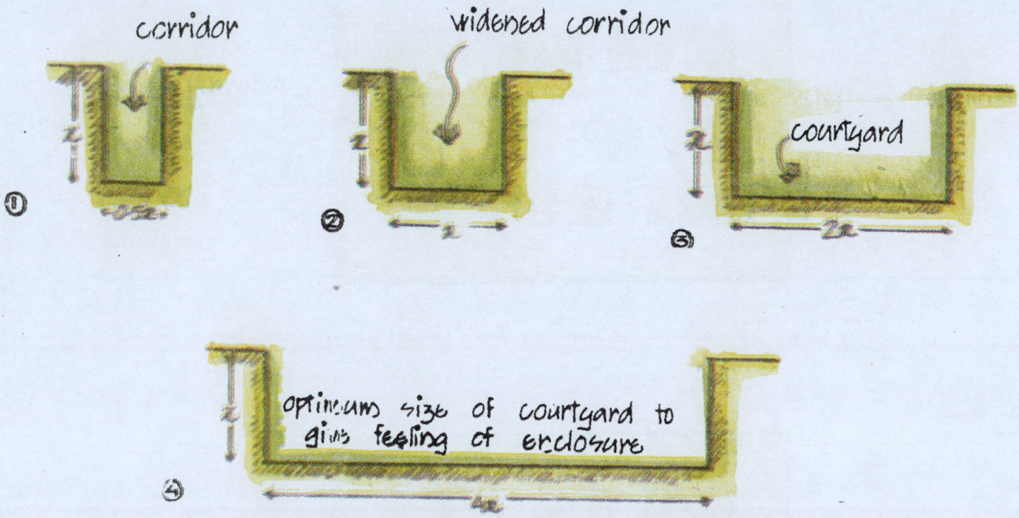

Fig. No. 26 : ***Ratio of Height of Building & Widths of Open Area***

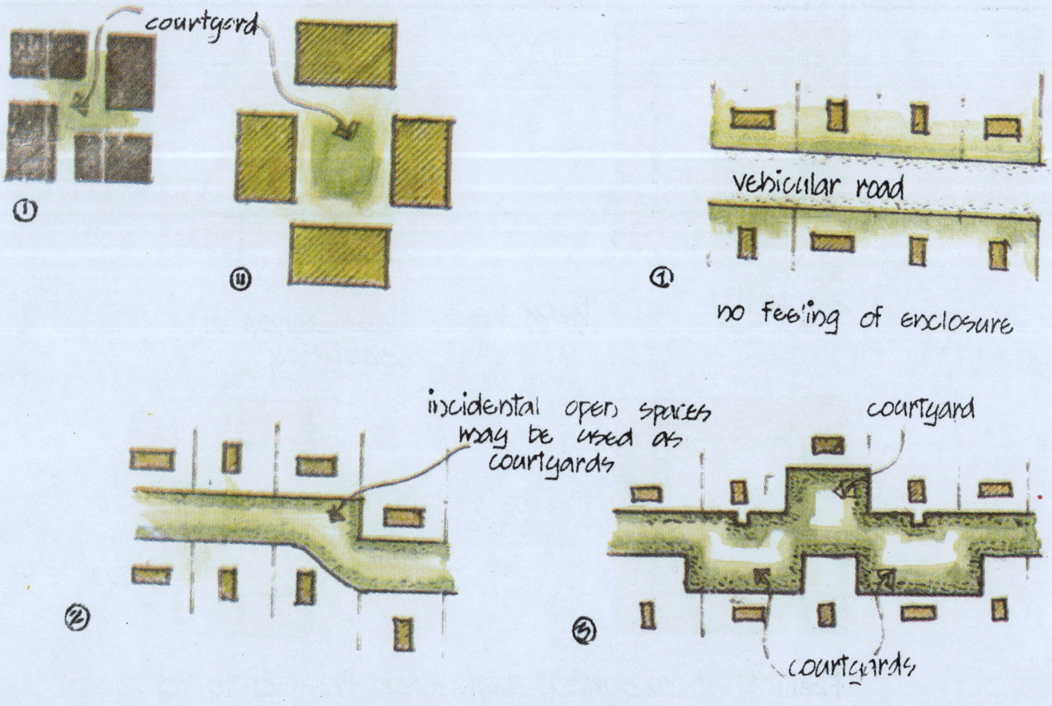

Fig. No. 27 : ***Types of Enclosures to form Courtyards***

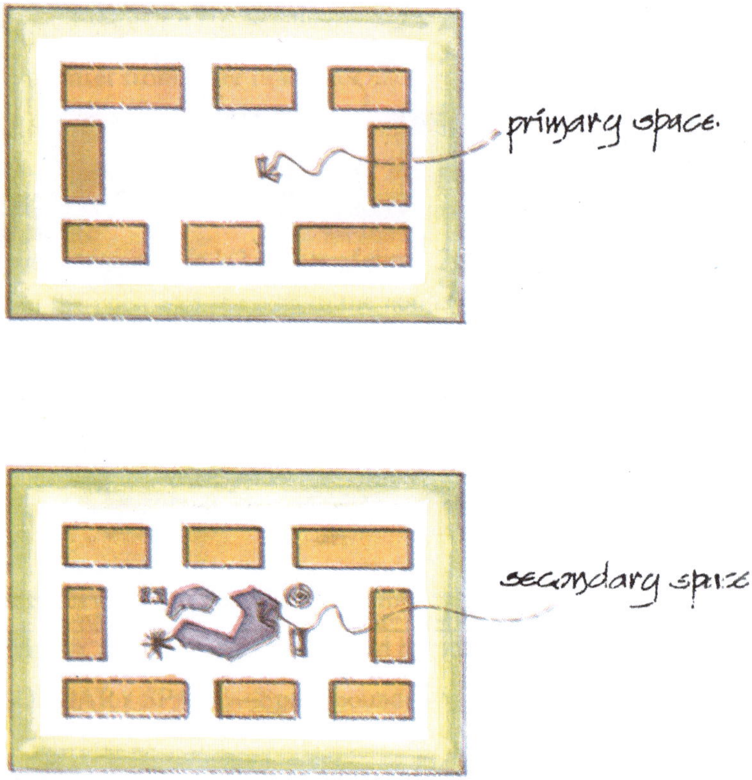

Fig. No. 28 : ***Spaces in Housing Layout***

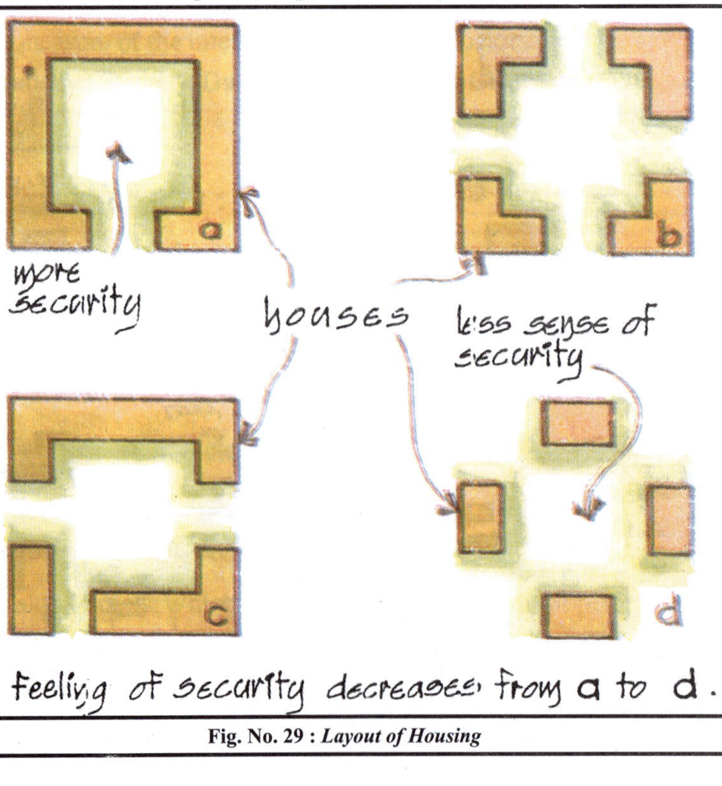

Fig. No. 29 : ***Layout of Housing***

(iii) Technical Factors : Like fire risk, Shortage of buildable land, access, ground conditions etc.

(iv) Economic Factors : Like Land cost, cost of transportation, cost of essential services.

5.6 BUILDING COST AND TECHNIQUES OF COST OPTIMIZATION :

Building Cost :

It has 4 major components:

(i) Cost of land and cost of land development.(Land filling etc.).

(ii) Cost of infrastructure like that of water supply, sanitation, roads and footpaths, electricity distribution and illumination, garbage disposal (solid waste management), landscaping and arboriculture, land cost for community facilities.

(iii) Building cost including materials (cost of construction of foundations, superstructure, roof etc.);and cost of labor (which can vary between 25%-30% of the building cost).

(iv) F.M.O.C.(i.e.Fees, Management, Overheads and Contingencies).

Optimization of Building cost :

This can be done by :

(i) Material Economy,that is, economy in material selection, standardization, and flexibility of building code.

(ii) Labor Economy, which can be got by avoiding labor intensive building work, timely supply of materials and equipments and adoption of management techniques like P.E.R.T. (Project Evaluation Review Techniques), C.P.M. (Critical Path Method) etc.

(iii) Design Considerations, like shape of plan (by reducing unnecessary notches in the walls), size of the buildings, number of stories in a building, by reduction of circulation space and by reducing the height of the stories.

6

NEIGHBORHOODS

The modern concepts of Town Planning is often looked as an 'investment' yielding 'returns'.But cities are for the people and the social inter-dependence amongst them form the 'matrix' of Urban existence and expansions. Thus Clarence Stein and Clarence A. Perry advocated the concept of 'neighborhood'.

It is defined as a physical environment in which a mother knows that a child will have no traffic streets to cross on his way to school, which is within easy walking distance from home. It is an environment in which the house-wife may have an easy walk to the shopping center where she may obtain the daily household goods and the man of the house may find convenient transportation to and from his work.

It is an environment in which a well equipped playground is located near the home where the children can play safely with their friends. The parents may not care to maintain intimate friendship with their neighbor, but the children are so inclined and they need the facilities of recreation for the healthy development for their minds and spirit.

Hence, in the concepts of Stein and Perry, the cities are viewed as social meeting place and not merely a collection of houses,workplaces,road networks and with the provisions of the physical infrastrucrures.The whole city is divided into several neighborhoods, each accommodating a considerable population (say 5000), and providing community facilities. Each neighborhood is segmented to accommodate different income groups. No high speed vehicles are allowed to cross the neighborhood but pass only from the peripheries.

Fig. no. 30 : AN IDEAL NEIGHBORHOOD

Fig. no. 31 : NEIGHBORHOOD MODELS OF STEIN AND PERRY

Fig. no. 32 : MODEL OF N.L.ENGELHARDT(JR); and of JOSE SERT

Jose Sert planned the neighborhood in a linear fashion, placing the industries at one side and the schools and the commercial centers at the middle of two rows of neighborhood units.

This 'module' could be repeated (in mirror image) to increase the length of the neighborhood units.

N. L. Engelhardt proposed his neighborhood unit strictly on the basis of the location of the schools and academic facilities. Nursery schools were places at locations catering to people within a radius of 1/4 of a mile. Elementary schools were having a coverage of people within a radius of 1/2 mile. Middle schools were placed to serve the people within a radius of 1 mile and that of upper school was 11/2 miles.

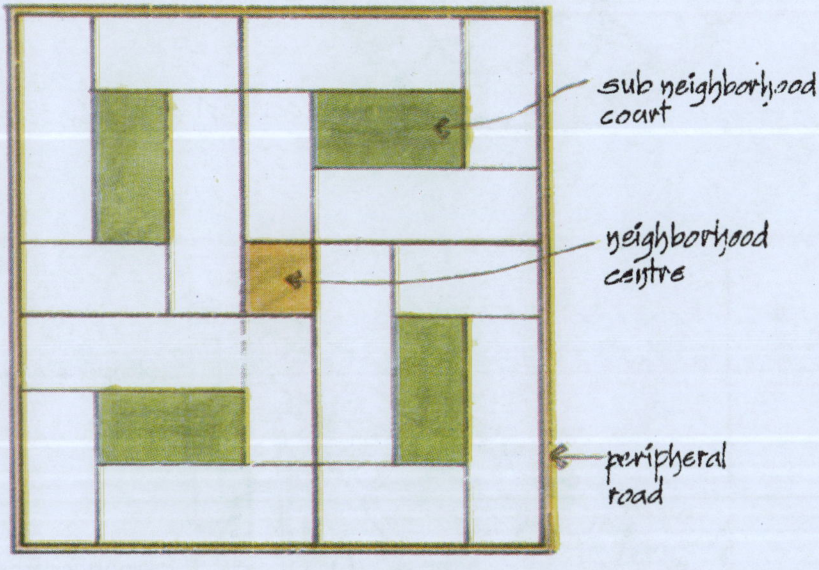

Fig. No. 30 : ***Ideal Neighborhood***

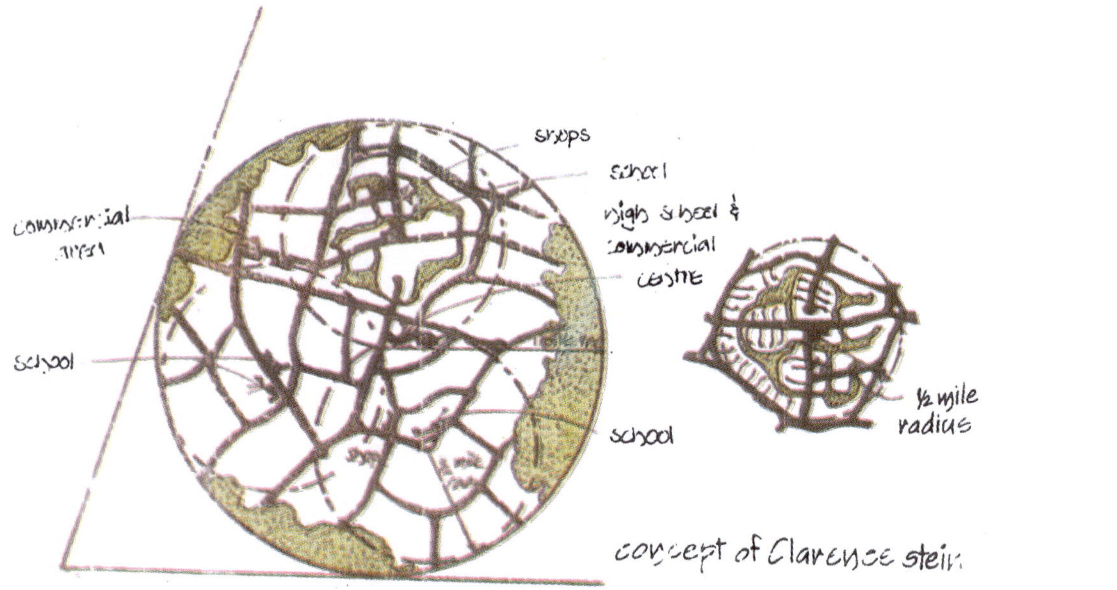

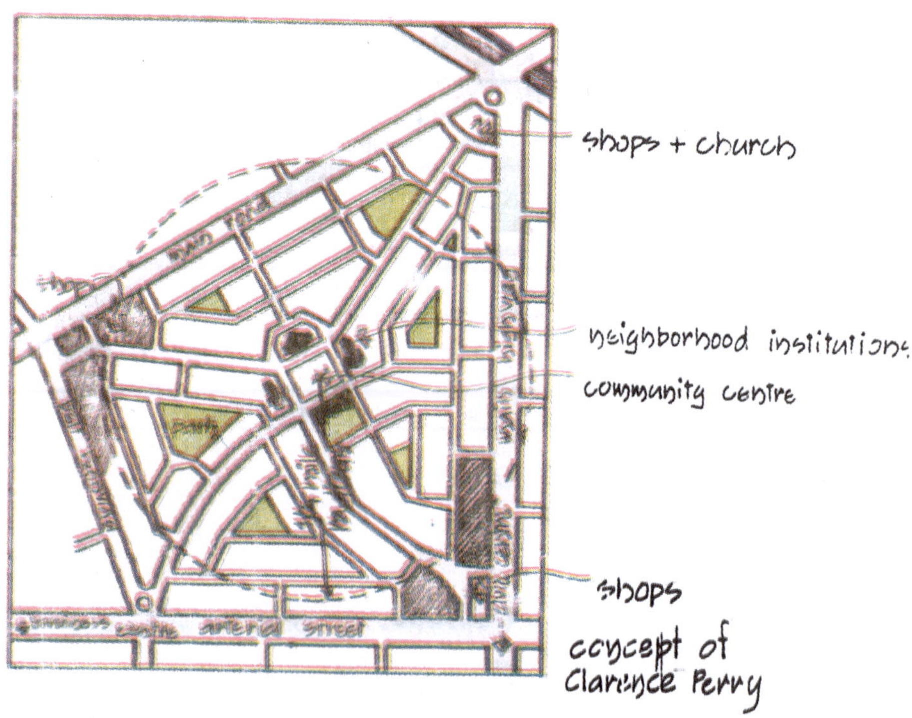

Fig. No. 31 : *Neighborhood Concept of Clarence A. Perry & Clarence Stein*

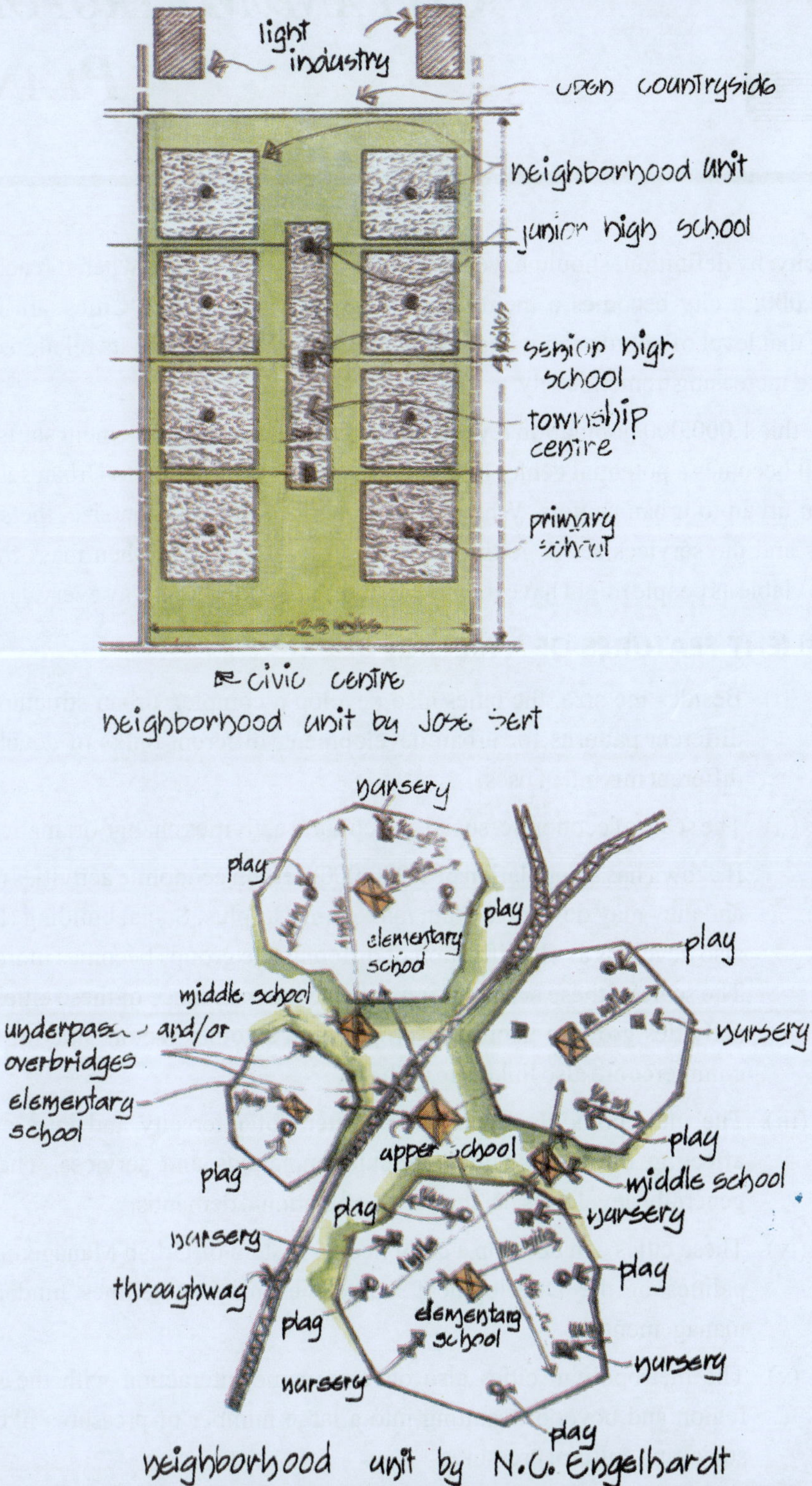

Fig. No. 32 : *Neighborhood Models of Jose Sert & N.L. Engelhardt*

7 CITY AND METROPOLITAN PLANNING

A city, by definition, should have a population of 100,000.; and when it reaches to a level of 1,000,000, a city becomes a metropolis. Study of 'Metropolitan Cities' are important as reaching that level of population it becomes a very complex structure its relationships with its hinterland increasing tremendously.

So, this 1,000,000 population level is a mark when an urban settlement starts growing on its own. It becomes a potential center of attraction not only from rural to Urban settlements but also from urban to urban centers. When a city reaches the 0.5 million size, the structure, the networks and the services starts getting complex. As for example, then mass transportation becomes viable as people might have traverse 10 to 15 kilometers one way every day.

7.1 SALIENT FEATURES OF A METROPOLIS :

(i) Besides the size, the cities also develop a complex urban structure displaying different patterns for urban development, different bulks of development and different mixing of uses.

(ii) The scale of economic, social and cultural activities change dramatically.

If a town has a population of 1,000,000, certain economic activities may develop and they may develop certain marketable surplus. Social building (like schools) also develops as classification of different age groups becomes more prominent. The scale of these activities indicate the independence of these cities. Economic activities grow in geometric progression as other factors like business, trade, commerce are also linked up with it.

(iii) The metropolis also develops a pattern of inter city and intra city linkages affecting the movement of population, goods and services. These linkages generally develop as per the various functional demands.

(iv) These cities can develop a complicated system of Urban Management. Municipalities or the Development Authorities may sometimes hinder the urban management.

(v) The metropolitan cities also develop some interaction with the surrounding region and beyond, resulting into a large number of pressures like economic, social and cultural pressures.

(vi) Development of the problems of the Inner City or Core Areas, which may not match with the new developments, but can be mitigated with one another.

(vii) Growing informal sector activities and unauthorized developments are seen in these cities. It is generally accepted that informal sectors are a disease of the metropolis and not of smaller towns.

7.2 INDIAN URBAN SCENE :

In 1951 in India we had only 5 metropolitan cities. In 1961 it was 7, in 1971 it rose to 9, in 1981 there were 12 and in 1991 there were 23 metropolitan cities. According to a projection done by the World Bank the total population by 2001 will be 1000 million of which 300 million will be urban. In 1991 the total urban population was 218 million which means that we will be adding nearly half the total population of Urban India in a matter of few years. Hence the urban areas need a specific policy about where the new population will grow, and which town will be given more incentives to grow. Therefore planning a city is related to the settlement pattern at the state level. The regional context of the city and the metropolitan area is very important, as a smaller city may have a functional specialization over a larger city; as for example, Roorkee University is more important-than the University of Meerut.

The urban areas are increasing in number and the existing urban areas are increasing in size, as a result the coverage of water supply and other facilities are reduced to a great extent. Larger the city size, more private investments are attracted as more returns are expected there (as in case of housing). The concentration of surplus wealth from the private sectors will flow into the larger areas.

In India, the Class II and Class III towns (having population of 50,000 to 99,999 and 20,000 to 49,999 respectively) are more in number than the other classes. This is because of migration. An in depth study of migration pattern show that the migration occurs in two stages. Villagers are attracted to Class III and Class IV towns as colleges, health facilities and job opportunities are more there, which are having a population range of 20,000 to 49,999 and 10,000 to 19,999 respectively. Class V and Class VI towns (having population of 5,000 to 9,999 and less than 5000 respectively) are semi-rural towns.

The Class III and Class IV towns do rot have any multiplying job opportunities, hence the first generation of the migrated villagers come and settle in Class I cities. Hence, when we say of development, promotion in all the aspects of the Class IV and Class V towns are out of question. The Fourth and the Fifth Five Year Plans started with the I.U.D.P.(Integrated Urban Development Program) and the Sixth Plan introduced the I.D.S.M.T. (Integrated Development of the Small and Medium Towns) schemes.

By these, Master Plans were made mandatory for all the cities. In I.U.D.P., for each Master Plan a core project was made which had overall development program. In the I.D.S.M.T: schemes, the potential of the Small and Medium sized towns were identified, which, with small incentives can develop into regional centers. These towns must have certain basic economic activity. Then only it can create more jobs and attract private developers. The settlement, to become a regional center, must have an approved Master Plan and 235 towns were identified in the Seventh Plan and 210 towns were taken up as there was problems of management and implementation. But by 1992, it was found that no proper monitoring was done towards the progress of the schemes of the Seventh Plan and the Eighth Plan also gives importance of the schemes adopted in the previous Plan. Other new areas of thrust were also identified in the Eighth Plan like the promotion of the non-agricultural activities, upgradation of the informal sector occupations and their integration with the urban fabric of the large towns and cities. It also emphasizes on the development of the infrastructural services in the urban areas, development of the urban housing, development of the urban transport system and the development of the weak financial and organizational base of the urban Local Bodies.

No city of India can be termed as a Primacy City of whole of India (Primacy Cities are those cities which have a major share of the development grants, both private and public.). Primacy is the establishment of 'Super Order'. Primacy is seen mostly at the State and Regional level. In India, metropolitan growth takes place in the shape of 'Urban Agglomeration'. So, censuses gives the population of Urban Area, within the boundary of which there may be more than one municipal boundaries.Each municipality have the power to make the plans and implement them and enforce the regulation from time to time.

The authorities of the municipalities and that of other planning and non-planning agencies remain, but still, there is a continuity of the development of the cities. Thus developments are not 'planned' ones. The problem can only be solved by a proper settlement planning at the sub-regional and regional levels.

APPROACHES TO DEVELOPMENT: The issues which engaged us in various kinds of debate.

We became conscious of development just before independence in 1945-55. At that time two types of ideologies were there. The so called developed countries developed a '**Guilt Complex**' of draining entire resources from the poorer countries. This 'Guilt Complex' gave rise to 'Sympathy Syndrome' which came in the form of aids in food, education, technology etc. Each of these came with a separate strain. The developed countries thus started training their personnel in the developing countries, for their own benefit. As for example, doctors and engineers were produced at are lower cost in India, and because of less job opportunities, they went to the developed countries and thus enrich them. This has been termed as 'Neo-Colonialism'. This was very clearly understood in the late' 50s. By the early '60s, it was thought that the Western developing model, known as 'Keneysian Model' of Harrod Dammor, must be adopted. This Model aims at the increase of the GDP (Gross Domestic Product) which is independent of the distribution system.

By adopting this, it was found that the disparities between the rich and the poor were increasing further. The 'survival of the fittest' were there, but the 'trickling down' system (i.e. the thought that the benefits of the development will 'trickle down' from the rich to the poorer sections of the society) did not work. More backward the region, more of these disparities came up. Development thus also means 'distribution' and not only growth. Between 1960s and 1970s people realized that an increment in the GDP or per capita income is not going to help. Therefore, three theories were accepted. These are:-

1. Agriculture should be protected and promoted — all ancillary institutions must be protected by subsidies and low taxes and promoted through science and technology.
2. Balanced regional development should be there.
3. Rural reconstruction must be done.

Thus, the first theory resulted into 'Green Revolution', the second theory lead to the adaptation of Regional Development in Planning and the centralized concept of planning was evident not to be Universal.

The rural reconstruction theory was advocated by U.S.A.; U.K. and France. None of these countries in the last 100 years have realised what village economy is. So our condition was inconceivable to them. The two main ideas of rural development that came up were :

1. Development from the top; that is, putting industries and imparting mass education to the rural areas.

In these schemes, the beneficiaries (i.e. the people) were not allowed to participate in the process. As a result the people of the village could not understand the package and they became dependent on the aids. They were not interested in developing their resources or boosting their agricultural base etc. The Rural Reconstruction Programs failed as the packages were conceived without the people. So, the regional dimension could not be interwoven in the rural situation. By the end of 1960s it was clear that industrial development at few spots cannot develop the region as a whole.

2. Around 1972, the 'Basic Minimum Needs' Program was started. By 1975, a question was developed like, WHOSE basic minimum needs are to be fulfilled and WHAT standards are to be taken to identify the basic needs.

Between 1970-75 it was thought that population control is the only way by which the fruits of developments can be equally distributed. The goal of development is — 'Welfare of Man', but the question remains, 'What is 'WELFARE' of 'MAN' ?

It was agreed upon that the imbalances in terms of well-beings are going to live with us and the overall balancing of disparities cannot be done away with. Hence, 'distribution' was given the most importance. This is also attached with the question of 'priority'. Each and every district may have a very different state of priorities; hence, priorities cannot be listed at National level.

Therefore, the question comes back, that development is something which cannot be given, but the people and Planners must work for it.

The major issues which engaged us in various debates are :

(i) GROWTH versus DISTRIBUTION — which one is to get more emphasis?

(ii) AGRICULTURE versus INDUSTRIAL DEVELOPMENT—which one should get priority?

(iii) URBAN versus RURAL Development.

(iv) CAPITAL versus LABOR INTENSIVE TECHNOLOGIES.

(v) CENTRALIZATION versus DE-CENTRALIZATION — what should be centralized and what should not be ?

(vi) MODERN versus TRADITIONAL.

(vii) SOCIO-ECONOMIC PLANNING versus PHYSICAL PLANNING.

7.3 APPROACHES TO DEVELOPMENT :

Basically there are three different approaches to development. They are :

1. INIQUITOUS INTERNAL ORDER :

Where development can be copied from the International scene, thus creating inequality amongst the various sectors of the society.

2. NEAR FEUDAL NATIONAL ORDER :

Where, after getting independence, when democracy started, all the power rested with the selected few. Thus the political climate is very much linked with the economic models; therefore the same people get selected over and over thus giving rise to feudalism. Democratic issues, as are carved out from the same channel, becomes dogmated.

3. ALIEN STYLE OF DEVELOPMENT :

Development means growth together with change; and the 'style of development' means the ways of doing things so that development takes place. There are differential levels of development in different sectors which have brought in different levels of change. This development have brought in some unhealthy traits in most of the less developed countries. The traits are :

(i) Inability to provide productive and remunerative employment to a large part of workers; as the rural areas cannot employ any more.

(ii) Inability to distribute the fruits of (economic) growth as per the likeness and needs of the people.The benefits are acquiring only to the very handful.

(iii) Inability to secure mass participation in the development decision making process; and thus the developmental consciousness is very less in these countries.

(iv) Social disruption and rising levels of violence of several kinds.

(v) Violation of the basic Human Rights by the group who hold power.

(vi) Squandering of irreplaceable natural resources and degradation of natural resources.

(vii) Unmanageable growth of population.

(viii) Over-concentration of investments and skilled manpower in metropolitan cities.

(ix) Lack of co-ordination between educational planning, health planning and social planning objectives within the framework of overall economic planning objectives.

The United Nations in 1985 gave seven criteria for evaluating the style of planning and development in any country. These are:-

1. Extent and nature of National Autonomy; that is, to what extent each nation^can take decisions of development, growth and distribution.

2. Extent and nature of Popular (People's) participation.

3. What specific directions are given regarding the sectors of production, production incentives, and the form of control over means of production; that is, if the public is free to sell the products in the open market.

4. Mechanism for redistribution.

5. Encouragement or otherwise to the form of (individual or collective) consumption of goods and services.

   For example, encouragement can be given to reduce the prices of motorcycles, but otherwise effect is a hike in the price of petrol, thus nullifying the effect of benefits.

6. Extent and nature of protection of human environment.

7. Extent and nature of protection of human relationships leading to solidarity, security, self-realization and freedom.

The above mentioned points are a set of ambiguous yardsticks by which the nature of development is measured in the eyes of U.N. All of these points are in a state of flux or constant redefinition.

7.4 THE MODIFIED PLANNING PROCESS :

The migration trends of all third world countries show that the migration is towards the metropolitan cities. About 33% of the migration takes place from rural to rural and the rest 47% constitute rural to urban, urban to rural and urban to urban migration. Of this 47%, 25% is from rural areas to Class I cities. Out of 159 million (1981 census) urban population, we had 27% contained in 12.metros the estimates of 2001 is 300 million. Now there are 23 metros and it is very likely that the Class I cities will grow not only in number but also in size:

When a city acquires the 200,000 mark of population, it looks for secondary centers of economic activities. This is the take-off stage and from this stage, depending upon the regional role it plays, a city grow at a very fast rate and it continues upto a population mark of, say, 2,000,000; after which it stabilizes a bit. The graph showing the rate of growth is shown in Fig. 7.4.1. The curve of growth is in the shape of a'S' and the time interval 'x' as shown is very important as, this is the time when the population is jumping and all the plans go hay ward. Past trends cannot give this trend of growth, and this time period can be anything from 10 to 30 years. This time is inevitable for a Class I city and the population and time both are very uncertain during this period. These trends are generally found in cities having a population of 100,000. These cities deserves a modified planning approach or planning process. During the time period of growth new areas have to be added and the problems created by the existing size of the city has to be taken care of.

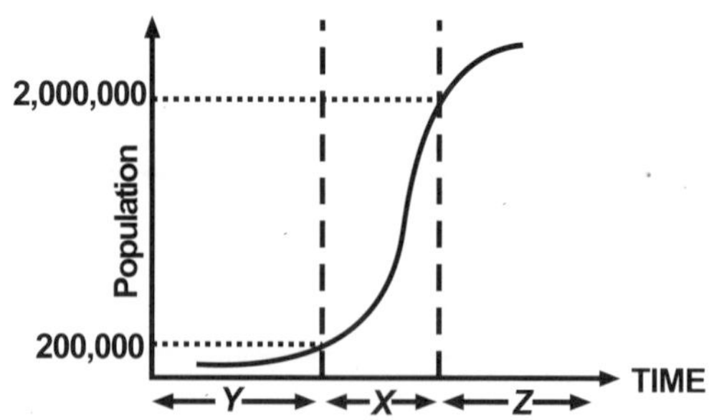

Fig. No. 7.4.1 : ***Development of Metrocities***

The lack of resources may be overcome by truncated actions but it may become difficult to co-relate the various actions after the time period. Thus this type of crisis to crisis planning process suffers from truncated decision areas. Now, a question arises, that there is no time for preparing a 'comprehensive plan' and if 'incremental planning' is done, the 'perspective plans' will be of no use. Therefore two things are important, one is a 'Long Term Perspective Plan' and the other is a 'Short Term Action Plan'.

The modified Planning Process is illustrated in Table 7.4.2.

It starts with the policy formation basing on the criteria of the existing Class I city.

The second stage is the Planning and Designing Phase which is arrived at by cumulating the present status and the recommendations for the future. At this stage a new set of norms and standards are also formed to be adopted in the Designing Phase. The Designing Phase is actually

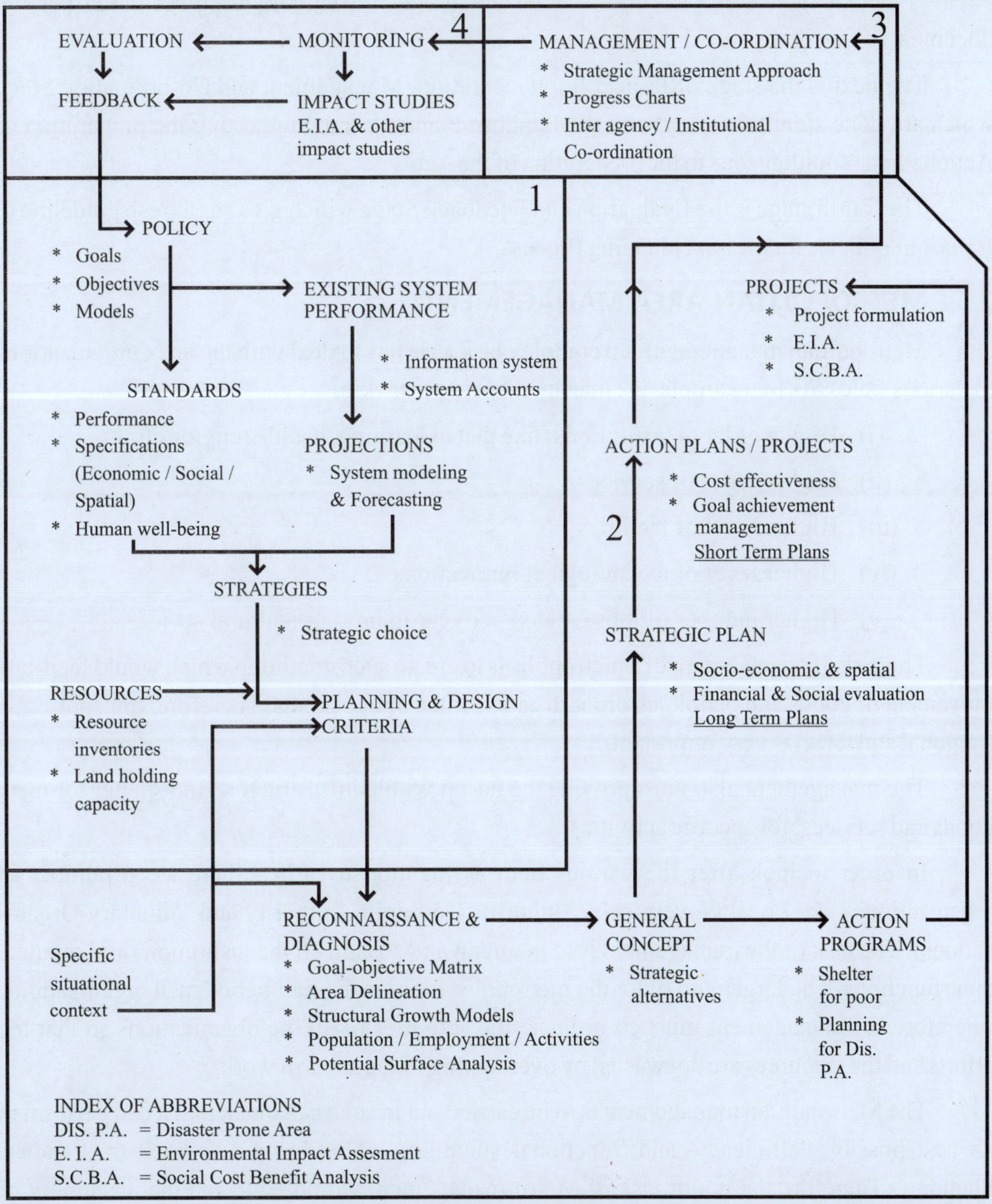

Table 7.4.2 : *A Model for Planning Process*

guided by the 'Structure Plan' covering much larger area than the Class I city in question to incorporate the decisions taken. The other guiding factor is the selection of the 'Action Areas'. Various 'Action Areas' could be selected to implement the decisions and of them, the Priority Action Areas are identified where the 'Instant Plans' are implemented as the situations of these areas worsens day by day. The resultant of the Structure Plan and the selection of the Action Areas is the emergence of a Local or Landuse Plan which eventually becomes the inspection document.

The next is the stage of Project Implementation, Management and Co-ordination Stage which are done simultaneously with the Landuse Planning thus suggesting the preparation of Action Plans; simultaneous to the preparation of the Actions.

The fourth stage is the Evaluation and Feedback Stage which gives us a fresh guideline to frame our policies for the next planning Process.

7.5 METROPOLITAN AREA MANAGEMENT:

Metropolitan management is a complex task as is has to deal with the apex organizations of various fields. As for example, the management has to tackle,

(i) Higher order of institutions (like that of learning, health, religion etc.)

(ii) Higher order of services.

(iii) Higher level of facilities.

(iv) Higher level of technological innovations.

(v) Higher number of jobs created per year in non-agricultural sector.

The task of management of a metropolis is to create such condition which would facilitate movement of goods and people, in order to cater for the above points. Therefore, transportation (regional and local) is very important.

The management also must provide the best possible infrastructure and shelter for men, goods and services; for specific activities.

In order to look after the various item as mentioned earlier, there are a number of organizations (like Local Government/Authority, Corporate Authority and Voluntary Organizations). The task of the management is to maintain and sustain all the institutions and regulate their functions. The larger and older the metropolis, more is the number of such organizations. Therefore the management must co-ordinate the activities of all the organizations so that the efforts and the resources are not wasted by overlapping of the areas of work.

The Metropolitan management is to be carried out in such a manner, that it can perform in its best possible 'efficiency' and 'functional suitability'. This bring one back to 'Landuse Planning'. Therefore the major task of Metro-management would be to run the machinery of Growth, sustenance and decay. These processes go on simultaneously thus making the whole thing much more complicated. Decay is a function of time, of which the result of man-made or natural calamity may not be stopped; but, decay due to neglect may be avoidable if management is done properly.

Any Management system works on the principle of accountability at various levels of organization. Here, the efficiency may be measured not only on the basis of quantity but also on the basis of quality and concept. The management system may not be effective if accountability in various levels of organization is not maintained.

7.6 OBJECTIVES OF METRO-MANAGEMENT :

1. To provide the people living in the city their basic needs, namely housing, infrastructure, community facilities and a desirable standard of living. This standard of living depends on the Cultural upbringing of the city which gives a metropolis its 'Image'— which is totally intangible.
2. To promote the economic activities which provide employment to the people living in the city, to ensure internal and external markets for the products produced in the city, and to promote the healthy growth of such activities.
3. To protect and conserve the natural amenities and resources in the metropolitan area and organize its sustained use without detriment to either the resources or the environment.
4. To organize the metropolis by developing the hinterland linkages, strengthening the economic, social and cultural ties that exist between them thereby enriching both the metropolis and the hinterland.
5. To promote and nurture the cultural, artistic and scientific activities along with technological innovations so as to make the metropolis a true image of the people and their culture.

7.7 ISSUES OF DEVELOPMENT MANAGEMENT SYSTEMS :

A Task Force appointed by the Planning Commission in January 1983 have sought out to relate various concepts and major issues of urban development management system as:

A. Assessment of the existing role of local authorities in the management of towns and cities and a critical review of their role and functions as expected and how they are discharged in practice.

B. Examination of the relationships of the Local Authorities and the Urban Development Authorities and their role in urban Management.

C. Indication of the direction in which the management of urban development to be improved and consideration of specific measures for maintenance of development assets and strengthening the local authorities.

D. Examination of the feasibility and desirability of the decentralization of urban management in large cities.

E. Examination of the existing system for training in urban management and suggestions for their improvement.

Thus, it becomes evident that for Metro-cities, 'De-centralization' is one of the keys to manage the city. The goal of the urban management should be to raise income and create employment opportunities and to provide the municipal and the other sectoral line agencies the necessary relief from their constraints of less organizational inertia, lack of financial and material resources and their inability to respond to the changes which takes place in time.

It may be said that the management of the urban development in India today is a real challenge to the planners and policy makers. It has several critical issues like the role of municipal bodies vis-a-vis the urban and metropolitan-development authorities, introduction of strategic planning, organization of metropolitan management, state level management functions and personal development for urban management. These issues along with the perennial problems of finances for urban development, constitute the critical area of urban development and its proper management. Some of the issues cannot be solved without taking a suitable policy decisions while others can be tackled by changes in management techniques, operational methods and organizational design.

8 PLANNING OF CENTRAL AREAS OF TOWNS

All the metropolitan cities of the world have very clearly defined 'old' parts forming the 'core' .These central areas or the 'core' of the 'old' cities which are known as 'Inner Cities'. The history of the town could be traced back to the old areas ; and these act as 'cultural anchors' of the society, the development of the cities and civilizations. Especially in the 20th. Century, after World War-I, the development of the metropolitan cities are so rapid that a difference between the building stock of the old and the new parts became very prominent. As per the definition sited by L.S.Bourne (1979), "The Inner City is defined as a geographic area or a set of areas in which the built environment, that is, the social overhead capital, is on the average older than that in the rest of the Urban Area and which, by definition occupies the central location.

Features of inner cities :

(a) There could be a number of detached parts of a metropolis which may be termed as Older Cities .These are the patches of traditional area still continuing,

(b) The core may grow and shoot out as per the theory of C.A. Doxiadis.

(c) The third possibility is where a new center grow in new area and start competing with the old centers having mixed landuse. But these does not qualify as Inner Cities as the built environment is made at a later date.The central location of the Inner Cities may not be a geographic center but have a function which is 'central'.

8.1 CAUSES OF THE FORMATION OF INNER CITIES :

(i) After the World War-I there was a hectic building activity in the western countries, which lead to the development of modern buildings in the fringes of the cities thus creating a contrast of the older and newer buildings.

(ii) The educated 'elite' classes moved out to the modern developments. Thus the Inner Cities became the place of agglomeration of jobless people, who are not very profitable taxpayers and hence all the municipal facilities and services were withdrawn from these parts. This resulted a declination of the property value and hence new investments were not incurred in the Inner Cities. Thus there was a deterioration of the living conditions of these areas. The rents were low, and hence an exploitative labor market was created. So, the percentage of the unemployed, old and infirm went on increasing in these parts. Here, deprivation was personalized. Slowly these areas became centers of illegal activities (like the centers of non-whites in America and England)and these areas are earmarked for social unrest.

8.1 CHARACTERISTICS OF INNER CITIES OF THE WEST :

(i) They have a declining economical base.

(ii) Overall loss of population.

(iii) In-migration of low income people.

(iv) Absolute Poverty.

(v) Existence of crime, disease, social pathologies—formation of ethnic and racial groups.

(vi) Physical deterioration.

(vii) Environmental pollution.

(viii) There is always fiscal problems due to declining tax-base.

(ix) Discrimination of private lending agencies and government housing programs.

8.3 THE CAUSES:

These characteristics are caused by :

(i) Aging and obsolescence.

(ii) Landuse competition.

(iii) Demographic Transition.

8.4 APPROACH TO STUDY/UNDERSTAND THE INNER CITIES :

There are mainly two approaches to study the Inner Cities to have a better understanding of the character of them. These are:-

(a) Problem Approach : The problem approach speaks about the results and the consequences of the nature of the Inner Cities.

(b) Process Approach : This approach talks about the processes behind the consequences. These may be social, economical, religious etc.

(a) Problem Approach:

The most common problems of the Inner Cities are :

(i) A declining economic base, loss of blue collar (industrial) jobs and high degree of unemployment.

(ii) Overall loss of population particularly in the productive age group. This particular phenomenon gives rise to lack of demand of higher order social services and lack of social diversity which is needed to keep the residential area attractive.

(iii) The in-migration of low income unskilled groups. The kind of jobs which are available are not in match with the skill of the migrated people and hence the unemployment is higher.

(iv) Absolute poverty exists because of disproportionate concentration of disadvantaged people in the Inner cities.

(v) Presence of crime, disease, social pathologies and fear of ethnic and racial groupings.

(vi) Physical deterioration of buildings, roads and social services such as schools, police, fire fighting, garbage collection etc.

(vii) Presence of Environmental pollutions.

(viii) Fiscal problems deriving from the interaction of the declining tax base and rising cost of social services.

(ix) The discrimination of the private lending agencies and government housing programs.

(b) Process Approach :

There are three basic setup processes. These are:

(i) The aging and obsolescence of the housing, social services, infrastructure and industrial base. Aging and obsolescence is the function of time and it is an inevitable process, but rejuvenation takes place during the time period. So, a whole area of a city cannot be three hundred years old as some part of it might be renovated ten years back. It depends on the economical health of the communities.

But if new migrants come and live in the older parts they do not have any psychological relationship with the old part of the city. So, the question of ownership comes. If the landuse of these areas changes, then construction or reconstruction may come out overnight by pulling down certain parts or making something new.

(ii) Landuse competition through the expansion of the central area or commercial functions as well as that of highways and industrial uses.

In the older parts of the Inner cities there are more pressures to change the landuse. This happens due to some speculators by whom pressurization of market forces take place. This occurs by commercial areas invading the residential areas. The commercial changes comes out at a lower cost level. No industries as such comes out which incorporates high cost, unless warehouses are taken as industrial use. Making Warehouse is the simplest type of conservation. If a highway pass through the inner City, it brings the fastest change along the road.

(iii) The demographic transition, notably the aging of the population and the loss of family household is another problem. If lone migrants, only single person households, are staying, then the social services of health and education will tend to disappear. This brings down the existing system and there is no fresh demand for these services. The aging of the population means lower order of taxation and they are not concerned with many things.

These three processes are highlighted as socio-economic constraints and realities of Inner Cities. Both the Problem and Process Approaches have one common thing to say, that the Inner Cities are declining (if they are really declining). But the question still remains, 'Why?'(If they are declining). To answer this, a series of hypotheses are given by various authors as stated hereunder :

(I) Natural Evolution Hypothesis :

Aging is a process of time but the core areas are the entry points of the in-migrants and the next of kin of the social groups can be easily located in older cities. Next to the core there is a transition zone of mixed landuses where rent is low. (Park's and Burgess Theory, 1925). This transition zone expands only horizontally. The social status and/or the economical status of an area is a function of the distance from the Central Business District. This has given rise to suburban development.

(II) Obsolescence Hypothesis:

The inner areas on a regular basis becomes the reservatory of obsolete technology, skills and building stocks. Obsolete, if replaced by the most modern things, may not be viable. The core areas have the intermediate or obsolete technology as they are available at a lower cost. Labor, in these areas are cheap. This hypothesis says that these obsolescence are to be maintained as more capitals can be generated by profit as there is competition in the market (as per Capitalistic Economical Theory).

(III) Preferences and Income (Pull) Hypothesis:

This is a preference coupled with social and economic opportunities. All opportunities move out as there is a 'pull' outwards and this is executed by the government in a selective manner. Only certain type of people who can establish new business with skills are pulled out.

(iv) The Unintended Policy Hypothesis :

When a new town is made the people from the Inner City move out (as seen in the case of London). So, the more disabled, aging, and infirm start surfacing in the Inner Cities. But de-congestion in the Inner Cities helps in-migration at the same time. So, de-congestion does not takes place but the people, who are able, move out. Hence negative factors like increase in population and unemployment increase in Inner Cities.

(v) The Exploitation Hypothesis :

This is similar to 'Structuralist' mode of analysis (Marx's Theory). All structurists were looking at a city as a consumption sector and the rural area as a production sector. As long as the mode of consumption is in the hands of few, then the advantages of capital will be shared by a few. Generating surplus will then mean accumulation of wealth (surplus) in the hands of few. This accumulation needs a machinery which will produce more and need least maintenance. The human beings can only be maintained at least cost, thus initiating exploitation. A large section of market is held at that location where cheap labor is constantly available and hence human machineries have been installed in the Inner Cities not by chance but by the interest of the capitalistic few. Informal sectors does grow in small towns as there is no capital accumulation.

This hypothesis is proved by many case studies. A study in Hong Kong University (1983) revealed that the policies that are created to rejuvenate the Inner Cities are deliberately made to get the availability of cheap labor, by forcing them to live as low in a standard of living as possible. Inner Cities grow, according to them, is by design and not by obsolescence, aging or decaying.

(vi) Structural Change Hypothesis :

There are two explanations of this — (i) Physical (incorporating material values) and (ii) Socio-Economic (incorporating human values). The social change is defined by the changes in the lifestyle, changes in the family structure and changes in the social relations. The economic part is defined by the changes in activity pattern and economical base.

For each one of them specific indicators can be found out. But these indicators are not universal and they changes from one country to another. But some social parts may have universal indicators as in the case of joint families. Decrease in the percentages in the number of joint families can be taken as the impact of modernization. Similarly the social relations are generally quoted by the effect of neighborliness (like shopping trips, lending and borrowing etc.).

Because of the removal of industries and certain other activities from the inner cities, certain other set of activities are generally taken up. These activities may be more incompatible.

But the question now comes, 'What is compatible?' The definition of compatibility changes over time — depending on the changes in the technology.

Changes in activity pattern have in most cases dwindled the opportunity of employment; because the skills of people are of no use as technological development have taken place. The change in the economic activity have been de-skilled by technology which has given rise to unemployment.[N]

Change in physical structure is defined by additions and alterations in the area; and this change in physical aspects are coming only to the buildings facing the roads. Therefore, the inner residential areas decreases as no incentive is got either from the private or from the government, sector, because of lack of service facilities. The physical change is also coupled with the removal of institutions from the old areas at least to the road sides, if not to far away places. Those who could pay taxes are removed and hence people living in Inner Cities are unable to pay tax and so services and facilities are cut off. Hence the condition of the Inner Cities are going down.

The structural changes have given rise to creation of two definite 'zones' or pockets in the Inner Cities. These are, one facing the streets and the other inner areas. Those facing the streets attract government and private investments and in some countries even the elevations of the buildings as 'heritage buildings' are preserved. But the much better buildings in the inner areas could not be preserved.

(vii) Fiscal Crisis and Underclass Hypothesis :

The problems of the Inner Cities can be seen as a 'Fiscal Crisis'. Preservation and maintenance of Inner Cities cause a problem in fiscal aspects. The recovery rate of these part are very slow and this lead the people to think that they will get reorganized by the individuals (market forces) as and when necessary. The cost of rejuvenation is almost twice than that of making a new city. Redevelopment by government is totally uneconomic. People also feel that the people of the Inner Cities and New Cities can not be integrated as the migrated people of the Inner Cities follow a life style of their country cousins. This is also a social liability.

(viii) The Conflict Hypothesis :

This deals with the ideological question or takes the ideological standpoint which goes on saying, 'Whatever you build today, will be an Inner City tomorrow'. So, Inner City is an inevitable part of a city and hence there is no need to delineate the region. Constant rejuvenation must take care of all over the city. If any old part comes down let it be rejuvenated by the market forces. New areas can create more jobs than Inner Cities. So there is no point of being sentimental. Whatever is done to maintain the old city, like all the old structures and heritage buildings could be retained. New usages has to be reconciled in the old buildings. But by that, neither the physical setting nor the cultural setting could be provided.

The second argument says, if redevelopment is needed, the face of the buildings can be retained and the interiors could be changes as done in Germany. That redevelopment is always, partial and half-hearted attempt to retain what happened in the past. But the question remains, 'How far is this justifiable?' This is yet to be answered.

Change in social, economic and physical aspects are taking place simultaneously . The Inner City areas must be kept active, and hence each Inner City is to be studied with respect to the role it plays in the metropolis.

8.5 INNER CITIES OF THE DEVELOPING COUNTRIES :

In the developing countries the scenario is different from that of the developed countries. Here the Inner Cities have old building stock thus containing the styles of architecture prevalent in a particular zone in a country. The richest of the businessmen are found to stay in these areas and the poorest of the poor, who migrated to the cities, also settled in these regions as the rents are low for the poor level of service, in comparison to the newly developed area. Everyday new investments (which may be unauthorized or illegal) are made in these regions due to the presence of the rich people lead to landuse change. Here the 'Inner Cities' are not dying propositions like the western countries and still they are spreading. This result in the development of mixed landuse both at vertical and horizontal level.

Here Inner Cities are defined not only by the area or the limits of the old city only but by a set of conditions as obtained in Inner Cities in terms of environment.

The Inner Cities serve as the core of the developing metropolis.They are not poor. The fringe of the inner Cities on a large scale is also vulnerable to these problems due to scarce resources. URBAN RENEWAL was the only weapon to tackle this problem.

The examples of renewal which we get are post war redevelopment that were bombarded during the war. When renewed the degree of coherence to the old city varies from place to place. In Warsaw, everything was rebuilt as a 'total replica' of what it was. In London, partial reconstruction took place — where the elevations of the old buildings were fit into the row houses which resulted in the development of 'hybrid Georgian' or 'hybrid Victorian' styles. This happened as there was a change in the building materials. After the second World War the economic regeneration activity was stepped up and on other side a reconciliation was going on in upholding human values and religious beliefs. Reconciliation is important to bring back the private investors in the government expenditure pattern. This resulted into Urban Renewal as the only method of dealing with problems of Inner Cities.

Urban Renewal was made an integrated part of the Development Plans.

But the Development Plans fail to undergo any renewal program as the tangible cost was much higher. So, 'SELECTIVE CONSERVATION' was done. The process of doing selective study was:

(i) SURVEY to analyze the Urban Structures and to identify the market forces behind them.

(ii) To identify the vital FUNCTIONAL AREAS; by identifying the landuse, density classifications etc.

(iii) To identify the CULTURAL HERITAGE 'ANCHORS'. Anchors are those institutions or those areas in which the people have faith.

(iv) To identify a set of criteria for HERITAGE EVALUATION —waiting too long for declaration of a building as a 'historical monument' may result in vandalism and deteriorating it esthetically.

(v) To integrate selected buildings/precincts/areas in the Development Plans.

It is a stupendous work of listing all structures on the basis of chronology. Then they are identified on the basis of historical values, present uses and the future utility of the buildings. This method have come to criticism in terms of their value-level approach.

vi) Cost and Benefit.

To get 'benefits', of the Government and private investments the following points must be borne in mind:

(a) The SOCIAL COST of renewal must be identified.

(b) Direct benefits must be got by reusing any buildings other than religious and to have PUBLIC PARTICIPATION. Thus, identification of the re-use value of the buildings, which may be put to some other use is the major task.

(c) Methods of plan preparation should also be different in these areas. Standards of access, parking etc. cannot be the same like the other areas. The engineering limitations of old buildings must be taken into consideration with a new set of bye-laws, and with modified administrative machinery.

8.6 METHODS OF PLAN PREPARATION FOR INNER CITIES :

Making of plans can be divided into three categories, namely:

(a) For New Settlements

(b) For Old Settlements and

(c) For Extension of Old Settlements.

Inner Cities has a hazy twilight area surrounding them having mixed characters; that is, there is a semblance of a boundary of an Inner City. Hence the major constrains that are faced while preparing the plans for Inner Cities are:

1. It has to be a defined area.

2. It may be surrounded by new developments either in all directions or in some directions. This new development has own pattern of form and its own pattern of landuse. Thus, there is a 'duality' in the area which is difficult to handle. Now, whether this variance is to be retained or changed, becomes a very important question.

3. To finalize the objectives of reconstruction. That the reconstruction should be done matching with the newer developments or whether the forms of the old buildings to be restored is the most important issue to be solved. Another important question is whether the street patterns to be changed or not.

4. To find out the mechanism to attract new investments. The new investments will come if facilities of bringing more services are incorporated. Now, it is difficult to incorporate new facilities in the Inner City areas. Non-traditional landuses may also come in pockets and it also becomes a challenge to determine the nature and extent of 'mixed landuses' that are to be incorporated. Mixed landuses are a phenomenon of age, time, need and money. All these are continually changing and hence there is no way to project the amount of mixed landuse.

5. Problems due to Rent Control Act. Rents are low in Inner Cities and removal of tenants is difficult in the Inner City areas. The number of Rent Control cases are increasing day by day. Hence nobody is interested to buy these buildings and the owners also are reluctant to built on their premises.

6. The financial conditions of the Local Bodies (Like Municipal Corporation) are very poor as revenue collection is very less. Hence the Local Bodies are not in a position to make plans or to implement the plans. Moreover, the Municipal bodies have old management systems and thus they fail to make or implement a plan.

Plan making in Old City acquires the job of a absolutely clear 'City Manager' who should manage these constraints and the 'Plan' will be one of his tools. Plans, legalities, organizations and management skills are the major aspects to develop the Inner Cities. The Plan required for Inner City would be from 'Policy to Layout' all embodied into one; i.e. a 'Perspective Plan'. There also should be a suitable authority to exercise the plan which should not follow the models of any Municipal Council or that of any Development Authority. This should be a process of regeneration with specially powered authority. In terms of a Structure Plan it is almost like a Regulation 8 Plan (i.e. making it a special area). Inner Cities will get the attention of the Local Authorities, on their own, over a period of time.

9 LEGISLATION IN PLANNING

Legislation forms a very important tool in the hands of Town Planners as the laws help the planners to develop the cities as they want. Without the laws it becomes impossible to implement the new concepts of planning and to restrict the developments which are not conducive for the general well fare of the city dwellers. The two most important laws that help the planners to implement their plans are:-

(i) Eminent Domain — by which the government can acquire any private property for the sake of public welfare, public health and public safety.

(ii) Police Power—by which the developmental authorities can control the nuisances and the undue developments. Most of the Developmental Controls like building bye-laws and planning laws emanate from this police power.

Nuisances are of two types, viz,

1. Private Nuisance, which affect the enjoyment of the individual, and
2. Public Nuisance, which affect the health, safety, moral values and comfort of the public at large. These nuisances come in the form of noise, air pollution, water pollution etc.

The police power takes care of these nuisances and also it empowers the authorities to look after the issues like architectural conservation, esthetics, spiritual values etc.

The police power is generally exercised by the municipal authorities who work with mainly two instruments like :

1. Building bye-laws, which establish municipal safeguards of the constructions and use of buildings.
2. Development Controls, which shape the overall Urban Network.

9.1 OBJECTIVES OF DEVELOPMENT CONTROLS :

The main objectives of the development controls are the following:-

1. To secure reasonable light, air and access to the buildings.
2. To secure facilities for sanitation, public health, safety and convenience.

3. To give protection against fire and structural safety.
4. To maintain architectural harmony in a locality.

The building bye-laws, the development regulations and rules are made to bear a relationship to these objectives.

9.2 COMPONENTS OF A GOOD DEVELOPMENT CONTROL SYSTEM :

For maintaining a good development the points mentioned below must be incorporated in the development control system.

1. There must be some obligation of the developers to seek permission of the developments.
2. There must be a clear concept and definition of'development'.
3. There must be a very clear mode, procedure and concept of seeking permission for development.
4. There must be uptodate set of norms basing on which the authority sanctions or denies permissions for development.
5. The authority must be empowered to refuse any development which not conforming to the norms.
6. There must be an appellant authority. (Though in India no such provisions are there).
7. The Local Authorities must be empowered to take penal actions against any developer who violate the developmental norms.

9.3 DRAWBACKS OF EXISTING DEVELOPMENT CONTROL SYSTEMS IN INDIA:

Most of the Indian cities could not exercise a proper control over the development as there are many inherent problems in the development control systems. The major problems identified in our development control systems are as follows:-

1. Most of the bye-laws are out dated.
2. The enforcement machinery is slack.
3. Public perception is bad for the development controls and they generally take these as a source of harassments.
4. There is always some frictions in between the politicians and the technical persons and this make the implementation of the developmental norms more slack and non-functional.
5. The people working in the Local authorities take the advantage of the slack administrative machinery and the friction between the politicians and the technical persons. Hence passing the plans of any developer becomes a source of

additional income to them. This makes the people to think the developmental norms as more harassing.

6. Even technical persons like the architects and the engineers feel that the bye-laws check their activities and hence they are not very keen to follow them very strictly.

9.4 TECHNICAL CONSIDERATIONS FOR FORMATION OF BUILDING BYE-LAWS:

To frame the building bye-laws, the Local Authorities must take into consideration the following points to make them functional and practical. These are:

1. Structural dimensions of walls and foundations.
2. Building height and road angles.
3. Front set backs for esthetics and road widening.
4. Density and F.S.I.
5. Bulk of buildings.
6. Urban Form.
7. Minimum dimensions of rooms, toilets and kitchens.
8. Side open spaces and the space in-between buildings.

9.5 PROCEDURE FOR OBTAINING PERMISSION OF A BUILDING :

The general procedure for obtaining the permission of construction of a building includes three steps, namely,

(i) Application

(ii) Scrutiny and

(iii) Sanction.

To get a permission of a building a payment is required in the name of 'building application fee' which covers the cost of site inspection, scrutiny and inspection during construction.

Apart from this, a 'development charge' is also taken by the Local Authorities. This is a one time charge to be paid before the initiation of construction of a building. This charge is levied on two grounds :

(i) For costing in the investment made by the government and

(ii) Cost for creating extra load on the infrastructures by the developments.

9.6 TYPES OF PERMISSIONS :

For any development the Local Authorities may give permissions which are of various types. These are:

1. CONDITIONAL PERMISSION : Where the Local Authority gives the permission for certain development but certain conditions are attached to it. The major

problem with these type of permissions is the people do not have any' intention to follow the conditions and the Local Authority sometimes also have no power to look after conditions laid against the developments.

2. PERMISSION BY RELAXATION OF BYE-LAWS : Where the permission is given by relaxing the norms of the existing bye-laws for a particular project. The major problem in this type of permissions are :

   Vested interests cause these relaxations and thus 'favoritism' may take place for any influential individual.

3. DEEMED PERMISSION : If within a certain period of time the authority does not give any judgment for any project the owner can carry on with the construction but only on certain conditions. These are:-

(a) The owner must wait for the time as defined in the statutory document.

(b) The development must not contravene any statutory bye-law of which the owner is not quite sure.

4. REVOKE PERMISSION : If the construction of any building takes place due to mis-interpretation of the building bye-laws, the Local authority may revoke the permission and the construction may be demolished.

9.7 REFUSAL OF PERMISSION :

Refusal of a permission can be done by the Local Authorities but it cannot be done on the ground that:

(a) Any public purpose proposed to be done in future on the same site.

(b) The building is not esthetically good. (Except where there is an Urban Arts Commission).

Refusal can be done only if there is a sanctioned statutory scheme.

9.8 OVERVIEW OF DEVELOPMENT CONTROL MACHINERY :

1. The development control machinery give Urban Planning a bad name and the bye-laws are extensively violated.
2. The bye-laws must be revised in every 10-15 years.
3. The space and safety standards specified by most of the Local Authorities are so lavish that only the H.I.G. can afford to maintain them.
4. Low density developments leads to Urban Sprawl.
5. The set backs are uniform everywhere in the city which sometimes seems illogical.

6. Very few Local Authorities have qualified planners and architects which lead to poor implementation of the norms and standards.
7. There are always delays and corruption in the Local Authorities which lead to unauthorized constructions.
8. Architects are not faithful to their professional ethics and they submit defective plans.
9. Very little attention is paid to Urban Conservation as there are no norms or act for conservation or preservation of buildings in the Municipal bye-laws.

10

URBAN LOCAL BODIES

There are two main organizations which look after the making and implementation of a plan of a city. These are the Development Authorities and the Local Bodies (i.e the Municipal Corporations, the Municipal Bodies etc.).

Where, the Development Authorities are responsible for the making of the Plans, it is the Local Bodies which are responsible for the implementation of it, along with maintenance of Public Health and safety. The various Local Bodies which are responsible for Urban Management are listed in Appendix III.

The major functions of the Urban Local Bodies may be listed as:

1. Public Health and sanitation.
2. Medical.
3. Public Works and Public Utilities.
4. Education and Social Welfare.
5. Development.
6. Administrative and General.
7. Miscellaneous.

Some the functions under each heading are obligatory, while some are discretionary.

A detailed study of the functions of the Local Bodies will help us to understand clearly the importance of them in the life of a city.

10.1 PUBLIC HEALTH AND SANITATION :

10.1.1 Obligatory Functions:

(a) Supply of safe and potable water and inspection and analysis of the piped water supply.

(b) Regulating or abating offensive and dangerous trades, calling or practices.

(c) Abating of all Public Nuisances.

(d) Public Vaccination and inoculations, prevention and control of contagious, infectious and dangerous diseases.

(e) Prevention of water and air pollution.

(f) Rodent control.

(g) Acquiring, maintaining and regulating places for the disposal of dead bodies. (Like crematory, cremation grounds, burial grounds and cremation of unclaimed dead bodies and carcasses of animals).

(h) Disposal of stray dogs and wild animals.

(i) School Health Programs.

(j) Clearing public spaces like roads etc.

(k) Collection and disposal of night soil and rubbish and preparation of compost manures.

(1) Constructing and maintaining public latrines and urinals,

(m) Construction and maintaining of drains, cleansing of drains, sewers etc.

(n) Prevention of food adulteration.

10.1.2 Discretionary Functions:

(a) Acquiring and assisting in the acquisition of suitable places for carrying on offensive and protected trades.

(b) Establishing and maintaining farms or factories for the disposal of sewage.

(c) Organization, maintenance and management of chemical or bacteriological laboratories for the detection of diseases, adulteration of food stuffs and drugs and research in the field of public health.

(d) Watering public streets and places.

10.2 MEDICAL :

(a) Establishing and maintenance of Public Hospitals, dispensaries, maternity, child welfare and family planning centers and ambulance service.

(b) Maintaining and assisting the maintenance of veterinary hospitals.

(c) Provision of anti-rabies treatment.

10.3 PUBLIC WORKS AND PUBLIC UTILITIES :

10.3.1 Obligatory Functions:

(a) Construction, maintenance, improvement and alteration of public streets, bridges, sub-ways, culverts, causeways etc.

(b) Control and regulation of Building activity.

(c) Lighting Public Streets, places and buildings.

(d) Planting and maintaining trees on road sides and other public places.

(e) Provision of housing for conservancy staff.

(f) Securing and removing dangerous buildings and places.

(g) Construction and maintenance of cattle pounds.

10.3.1 Discretionary Functions :

(a) Supply of electricity and gas.

(b) City transport services.

(c) Milk supply schemes and setting up of dairies, poultry etc.

10.4 EDUCATION AND SOCIAL WELFARE :

10.4.1 Obligatory Function :

(a) Establish and maintaining pre-primary and primary schools.

10.4.2 Discretionary Functions :

(a) Establishing, maintaining and assisting institutions of secondary and higher education.

(b) Undertaking cultural activities and assisting cultural institutions.

(c) Establishing, maintaining and assisting institutions of physical culture.

(d) Undertaking measures and programs for social and normal welfare of the citizens.

(e) Provisions for milk or mid-day meals for school children.

(f) Undertaking measures for and assisting adult literacy and social education.

(g) Undertaking Urban Community Development Programs and promoting public participation.

(h) Organizing, maintenance and assistance to institutions for infirm, sick or incurable persons.

(i) Supply milk to expected or nursing mothers and infants.

(j) Providing and assisting in the development of homes for orphans and destitutes,

(k) Construction, establishing and maintaining libraries, museums, community halls, swimming pools, places for entertainment and recreation etc.

10.5 DEVELOPMENT:

10.5.1 Obligatory Function:

(a) Construction, altering and maintaining markets, shopping centers, slaughter houses, baths, washing places, drinking water stand posts, wells, public parks, gardens, water troughs for cattle and maintaining tanks and grazing grounds.

(b) Constructing and maintaining parking places and vehicle stands.

(c) Improving agriculture including reclamation of wastelands in the rural pockets inside Municipal limits.

10.5.2 Discretionary Functions:

(a) Development of building sites, construction of houses and their disposal.

(b) Encouraging formation of housing co-operative societies and giving assistance to such societies:

(c) Undertaking schemes of slum clearance and programs for redevelopment.

(d) Provision for suitable accommodation for cows, calves etc.

10.6 ADMINISTRATIVE AND GENERAL :

10.6.1 Obligatory Function :

(a) Preparation of annual and other reports.

(b) Maintenance and development of municipal properties.

(c) Naming and numbering of streets, public places and premises.

(d) Erecting substantial boundary marks defining the limits of the municipal area.

(e) Regulation of traffic and provision of traffic signs.

(f) Removal of obstructions, encroachments and projections from the public streets, public spaces and public places.

(g) Registering the births and deaths and other vital statistics.

(h) Protecting life and properties from fire hazards and making arrangements for extinguishing fires.

(i) Fulfilling obligations imposed by law and enforcement of rules, bye-laws etc.

(j) Promoting well being of municipal employees and their dependents.

10.6.2 Discretionary Functions:

(a) Contributing to and participating in the activities of the local government institutes, training centers and associations of local bodies.

(b) Arranging public receptions, ceremonies, entertainment, sports etc.

(c) Arranging and controlling fairs, exhibitions etc.

(d) Construction of and regulating the use of lodging houses, camping grounds, tourist bureau and rest houses.

(e) Establishment and maintenance of printing press and workshop for municipal private work, if possible.

(f) Taking measure to control beggars and vagrants and providing means for their relief.

(g) Undertaking commercial and industrial activities.

(h) Conducting censuses and surveys.

10.7 MISCELLANEOUS FUNCTIONS :

10.7.1 Discretionary Functions:

These are mostly

(a) Establishing and maintaining nurseries for trees, plants and vegetables.

(b) Construction of and maintaining granaries, godowns, cold storages, ware houses and sites for preservation of food grains, food stuffs and vegetables.

(c) Regulation of weights and measures.

The Local functions differently in the different states of India. In some states it also acts like a Development Authority, or in some states it just works for implementation of the Plans and maintenance of the safety and security of the city.

In Maharashtra, some of the Local Bodies functions also as a Planning and Development Authority.

In states like Punjab, Bihar and Orissa, the Local Body and an Improvement Trust act as-the Planning and Development Authority with jurisdiction over the same areas as of the Local Body.

In Karnataka, Andhra Pradesh and Gujarat, a few of the Local Bodies and the Development Authorities act together having a jurisdiction larger than that of the Local Body.

In Haryana, The Local Body and a Planning and Development Authority act at the State level directly dealing with Planning and Development.

In Maharashtra, Gujarat and West Bengal the Local Body and a Regional/Area Planning and Development Authority act together having jurisdiction over local body area and also the intervening rural spaces.

In Madhya Pradesh and Tamil Nadu, the Local Bodies and the Planning and Development Authorities work with specialized agencies like Water Boards, Housing Boards, Slum Clearance Boards etc. all functioning in the area.

In Chandigarh U.T.; Area General Administrative setup exercise the function of the Local Body and the Planning and the Development Authority and specialized functional agencies.

A detailed list of the function of the Municipalities of the various States of India in attached in Appendix V.

Unit III

1 RURAL PLANNING

India is a land of villages. 74% of the total population is rural pf which 30% live below the poverty line. Because of the low Q.O.L.(Quality of Life) and practically no economical development, rural people migrate to the urban centers causing a great pressure on the cities. Main objective of the rural planning is to improve the housing sector and to search for an alternative source of employment of the village people.

The objectives to improve the housing sector are:

(i) To enable the villagers to lead a clean and healthy life.

(ii) Improvement of the housing will have a direct impact on the economy of the rural areas — more healthy environment will lead to a consciousness amongst the villagers to live better —they will exert more effort towards improvement of their agriculture and industrial productions; which will again provide them with the additional resources to meet their growing demands for better living.

(iii) Improved living condition will slow down the migration of the population from the rural to the urban areas.

The following table will give us an idea of the Indian Rural Scene;

Table — 01

Year	Rural Population	Urban Population
1961	82%	18%
1971	80.9%	19.1%
1981	76.69%	23.31%
1991	74.30%	25.70%

In 1971, in India, there were 439 million people living in the 5.67 lakh villages. The total number of households were 78 million, and the existing housing stock was 74 million wherein even thatched houses made of perishable materials and provided with a cooking corner were included. The unserviceable 'kuttcha' houses were 8 million, serviceable houses were 24.4 million (32%), semi-pucca houses were 27.5 million (37.2%),and 'pucca' houses were 14.1 million.

If these are looked from the international standards (U.N.) points of view, the condition is even worse. Speaking of water sources, wells were available to 68.08% of the households. Tanks and ponds are available to 6.45% of the households, tube-wells were available to 11.15% of the households and protected (piped) water supply was available to only to 2.56% of the households. 1.5 lakh villages out of 5.67 lakhs were earmarked as **problem villages** where there is no water within 1.6Km radius or upto a depth of 16m.

Only 4.5 lakh villages were connected by **all weather roads,** which gives access throughout the year.

3.5 lakh villages (i.e.60%) were still needed to be electrified, 97% of the houses had no toilet facilities and 96% of the houses had no individual bathrooms.

Thus to do away with the problems of the villages the Government has taken various means in the various Five Year Plans.

During the first three Five Year Plans , it was thought that the programs of development would raise the living condition of the poor in a 'trickle down' process. That agricultural growth without any major institutional reform will alleviate poverty. **Community Development Programs(1952), Intensive Agricultural District Program (I.A.D.P., 1964), and High Yielding Varieties Program (H.Y.V.P., 1966-67)** were started to increase agricultural production.

The Fourth Five Year Plan (1969-74) thrusted a direct attack on poverty namely through **Self Employment Programs** and **Wage Employment Programs.**

The Fifth Five Year Plan (1974-79) targeted to implement intensive agriculture, multiple cropping, application of H.Y.V. seeds, fertilizers, minor irrigation etc. for the identified beneficiaries with the special emphasis to dry farming practices and better management techniques. The plan also started the **Minimum Needs Program (M.N.P.),** to provide elementary education,rural health,drinking water provisions on housing sites, slum improvements, rural roads and rural electrification.

A revised M.N.P. was also made to give adult education, rural water supply, housing for landless laborers, and environmental improvements of urban slums and nutrition was also added to the previous list.

During Sixth Five Year Plan (1980-1985) adult education was added. Integrated Rural Development Programs (I.R.D.P.) is the central pillar of poverty eradication schemes. It was based on the detailed knowledge of the local needs, resource endowment and potentialities. The focus was on the maximum utilization of locally available resources, including local manpower, rather than on the introduction of large inputs from outside. The focus of I.R.D .P. was the target group comprising small and marginal farmers, agricultural laborers, and rural artisans. It aimed at integrating field programs to achieve it by developing the primary, secondary and tertiary sectors.

During the Seventh Five Year Plan (1985-1990), development on rural domestic energy, rural sanitation and public distribution system was added to the list. It enunciated land reforms to be an intrinsic part of the anti poverty strategy. The **Integrated Rural Energy Program (I.R.E.P.)** was taken up during this period. **District Rural Development Agencies (D.R.D.A.)** were also formed to review the existing administrative arrangements.

During the Eighth Five Year Plan (1992-1997) the land reform policies and I.R.E.P. schemes were modified to overcome the drawbacks of the previous plans. For Land Reform Policies five main principles were taken into consideration i.e.abolition of intermidiateries, tenancy reforms with security to actual cultivators, redistribution of surplus ceiling land. consolidation of holdings and updating of land records.

For I.R.E.P.,The Eighth Five Year Plan focussed at:

(a) Provision of energy for meeting the basic needs of working, heating and lighting; specially for the weaker section, by utilizing locally available energy resources to the extent possible.

(b) Provision of energy as a critical input in the economic development of the rural areas which result in the creation of employment, increasing productivity and income, accelerating the process of decentralized development.

The Plan emphasizes the need for integrated approach to rural development where the **Poverty Alleviation Programs,** the **Area Development Programs** and **Sectoral Schemes** are integrated.The Plan will take into account the physical and human endowments of the rural area, the felt needs of the people and the funds available.Viable activities, particularly in the agriculture and allied activities including animal husbandry, pisciculture, horticulture, forestry and agro processing would have to be selected. In addition, development of infrastructural support and backward and forward linkages will have to be ensured which are essential pre-requisites for the validity of the selected activities.

Greater self help of the people and their participation of the programs of the 'Panchyati Raj'.institutions, co-operatives and other self managed institutions have been sought in the planning and the implementation of the rural development programs.

1.1 PROBLEMS OF RURAL HOUSING :

There are basically 3 basic problems of Rural Housing, namely:

(i) Physical; (ii) Social and (iii) Economical.

(i) The Physical Problems:

The physical problems again are of two types, namely, ***At Dwelling Unit Level and At Village Level.***

Problems at Dwelling Unit Level :

Rain water plays a very damaging role in the deterioration of the village houses by washing out the mud walls of the structures. A probable solution to protect this, is to make the walls water proof. Proper foundation of these wall could be made with burnt bricks which could be continued like dado in the walls to protect the lower portions of the walls. Proper techniques of water supply and drainage could be made to improve the hygienic conditions. The toilet and the kitchen must be properly arranged. The other major problem is the number of rooms as the occupancy rate is very high in the villages. There is a lack of differentiation of the rooms within the dwelling units and most of the rooms also lack in proper ventilation.

Problems at Village Level :

Most of the Indian villages have an organic growth. The street pattern is narrow and crooked which does not match with the invasion of the new technologies like the scooters and tractors. The plots are of irregular size. There is also a lack of community facilities like health care, primary schools and provisions for water drainage and garbage disposal systems. The villages are needed to be connected with All Weather Roads so that whatever produce they have, they could be easily marketed.

(ii) Social Problems

Education must be given to the villagers agriculture being the first priority. The habits and practices which are having no logic what-so-ever are needed to be changed. Interdependencies amongst individuals create social exploitations. The farmers and the laborers who own agricultural lands are given plots in the 'Abadi' area (Residential Areas). The landless laborers do not have land in the village, and they cannot have land in the residential areas. So they are allowed to stay as long as they are serving the farmers after which they have nowhere to go. The government have initiated a number of schemes to over come these problems.

(iii) Economic Problems

The land is the most important aspect in the rural areas .The land holdings of the individuals are reduced when the land is distributed amongst the different members after a family is dissolved.The other problems are the lack of government help and subsidies for the various rural activities. There is also a tremendous lack of proper marketing facilities of the products and hence, the lions' share of the total produce of the villages remains unutilized thus rendering the villagers poorer.

1.2 AREAS OF DEVELOPMENT :

Certain areas are chalked out for the development of the rural areas of India and the government has taken many a steps for the overall improvement of the Quality of Life (Q.O.L.), of the rural population.

These areas are (a) houses; (b) Proper streets and drains, (c) Safe drinking water; (d) Open spaces; (e) Playgrounds; (f) Community Centers.

To ameliorate these problems the government has taken certain undertakings which are listed hereunder:

(A) Introduction of Village Housing project :

(i) AIDED SELF HELP : Bare minimum financial assistance and free of cost adequate technical guidance is given.

(ii) An all round economic development of the rural areas which enables the villagers to maintain their houses and to repay the loans taken by them for building their houses, without any undue hardship.

(iii) Improvement of housing condition in a village with the physical reconstruction of a village which are to be treated as a long term program spread over a period of say 10 years as to keep in step with the improvement of the village.

(B) Formation of Rural Housing Cells :

Village Housing Project Schemes provides for the establishment of the Rural Housing Cells to function as:

(i) To arrange physical (engineering) and socio-economic surveys of some selected villages.

(ii) To draw up the Master Plan for the selected villages on the basis of the survey data.

(iii) To prepare suitable design and specification of houses having due regard to local conditions such as climate, living habits, customs, local materials etc.

(iv) To provide overall technical guidance to local authorities in execution of projects under the schemes; and

(v) To generally ensure that money available for the purpose is properly spent.

(C) Rural Housing Wings :

This was made for research training and extension programs for the success of Rural Housing Schemes. The functions of Rural Housing Wings are:

(i) To impart training to the officers responsible for implementation of Village Housing Project schemes.

(ii) To conduct research related to the locally available building materials, techniques of construction, design of houses and village planning.

(iii) To carry out extension and demonstration work in the field for trying out and demonstrating the results of research and new construction techniques.

1.3 SOCIO-ECONOMIC ASPECTS OF HOUSING :

Objectives of Socio-economic Surveys :

(i) Assessment of peoples' demand for housing in the context of their overall demand pattern.

(ii) Assessment of peoples' resources, income and expenditure, their indebtedness etc.with a view to estimate their financial capacity.

(iii) Study of the present population according to age, sex, marital status etc. and its growth in a historical perspective with a view to project the future trends of growth so as to estimate the housing requirements including civic, cultural, recreational and educational amenities making allowance for future as well.

(iv) Assessment of peoples' attitude, their habits, customs.preferences etc.in relation to their impact in housing and planning.

Social aspects → Human beings → Mental hygiene → Education.

The physical surroundings have an impact on the Human beings and which in turn effects their mental hygiene.Hence to 'create' good environment the minds of the people must be made 'clean' through proper education.

The villagers should understand the necessity of a cleaner and healthier life, which would consequently result in better housing and ultimately better economical conditions. Hence the entire program will have to be brought 'home' to the rural public through intensive publicity based on visual and other aids.

RURAL HOUSING CO-OPERATIVES :

The co-operative societies can function in several directions :

(a) To avail loans from Government and distribute it to its members.

(b) To impart technical guidance.

(c) To evoke team spirit amongst the villagers to built houses.

(d) To contribute to the training of the villagers to change some of their attitudes in a manner favorable to improve production and enhance welfare as a whole.

AIDED SELF-HELP HOUSING :

The main theme is to combine the manpower of the families staying in the house, with the aid from the community (such as local materials, new techniques etc.). This would create a 'desire' amongst the people for 'improvement'. Hence, there is a felt necessity of convincing and educating people by some 'leaders' (like priests, chiefs, political leaders etc.).

Varieties of Self-help :

(a) The building of one room core structure by skilled labor and of the other rooms later by 'self help ' in slum clearance and resettlement projects.

(b) The 'self help' production of materials and building elements such as sand, building blocks, bricks, and lumber which are supplied to paid workers who build the houses.

(c) The foundation and the walls of the buildings are made by self help, but the roof is made by skilled labor.

1.4 FACTORS AFFECTING RURAL HOUSING :

1. Existing socio-economic conditions and the future requirements.
2. Developments taking place in the region (proposals for Regional Planning).
3. Influences of cities on sub-urban villages.
4. Extension of transportation facilities.
5. Increasing the level of agricultural output by means of mechanical farming.That is by tube-wells, tractors etc.
6. Influence of electric power.
7. Developmental activities under the Community Development Programs.

1.5 SURVEYS IN VILLAGE PLANNING :

1. Physical (plain table) survey.
2. Contour survey for roads, canals streams etc.
3. Detailed information about the sources of water supply, number and locations of wells, and disposal of human and animal excreta.
4. The maintenance data like : (a) Rainfall, (b) Temperature, (c) Climate (humid/ arid), (d) Nature of soil, (e) Suitability of local soil for (i) Making bricks,

   (ii) Constructing mud walls. (0 Available local materials, (g) Availability of skilled labor, masons, carpenters, blacksmiths etc., and their wages.

1.6 HOUSING PATTERNS :

Housing should be preferably low rise as land is not scarce and costly. Provisions for keeping cattle (either in the courtyard or in the back service lane) is to be corisidered. Density desirable is 25 to 38 houses per hectare. Sprawling village and thinner density lead to increase in the cost of roads , drains, electric lines etc. The desirable sizes of plots are 24m. to 27m. in depth. For semi-detached houses plot size like 9m. X 24m. and for row houses 9m. X 18m. plot are found to be adequate. Open spaces are required for children to play and for common purposes like social gatherings, fares, 'meals' ,etc.(Recommended areas for these purposes are 1.2 ha/200 houses). Space for manures and compost pits are to be provided at the fringe of the village keeping in view the wind direction so that no foul smell disturbs the villagers.

Road widths : Roads are to be such made as to accommodate two loaded bullock carts to cross each other without any hindrance.A minimum of 4.5m. for local streets and 9m. for streets passing through shopping centers or carrying through traffic to main streets and surrounding villages, are recommended.Too wide a road is also a handicap as it is very costly.

1.7 REDEVELOPMENT OF EXISTING VILLAGES :

Steps to be followed for re-development of the existing villages:-

1. A socio-economic along with a physical survey is to be done.
2. Planning standards are to be reduced to make the plan less expensive.
3. Provisions to be made for means to provide additional sources of income to the villagers in order to strengthen the economic possibilities for the work of reconstruction. The following flow chart gives an idea of the process to be followed in re-development of existing villages.

Village layout → Sector-wise planning → Selection of priorities like, drains/ roads/houses community facilities etc.

Unit IV

GLOSSARY OF TERMS

Base Maps : Base maps are done for Urban Areas where no good Planning Base maps are available or where none of the reference maps are completely satisfactory for city planning. There are various types of Base Maps, like,

(i) General purpose maps.

(ii) Detailed planning base maps.

(i) **General purpose maps :**

This map is frequently published for a city. It shows streets.corporation limit, the rail road, major public buildings, rivers and ponds. This map is used for generalized studies, records and displays for rough work. Generally, three sizes are there. These are:

(i) Wall Display size.

(ii) Table size.

(iii) Letter size.

(ii) **Detailed planning base maps :**

This map is prepared in an atlas form usually to the same scale as the topographic map in conjunction with a contour map. The atlas map may be keyed to the same grid system employed in the atlas of topographic sheets or it may be composed in terms of Planning Districts or actual Planning Areas. This map is in two or three series. The first series shows the street, right of way lines, rail roads, water course, lakes and Civil Division lines.

The second series shows the property and the easement lines combining with the series mentioned previously. In the third series, the structures are also added.

Boulevards

These are connections between large and medium sized parks. They permit people to drive in carriages through them and enjoy the surroundings. They may be called city street with good landscape and rural embellishments. Generally ornamental buildings face this boulevards.

Building :
(as per Delhi Municipal Corporation Act; 1957)

A house, outhouse, stable, latrine, urinal, shed, hut, wall (other than boundary wall), or any other structure whether of masonry, mud, brick, wood, metal or other materials (but not a portable shelter) is a building.

Central Business Districts

The C.B.D.(or the central Business Districts) is comprised of mostly landuses of commercial nature like shops for seasonal needs, commercial recreational centers like cinema halls, theaters, auditoria etc.

Roads from all parts of the city should give access to the C.B.D. The C.B .D. of a metropolitan city cater for a larger areas in the sense that satellite towns are also dependent on the said C.B.D. Hence easy traffic movements to the C.B.D. from within the city limits are required and also from the city fringe as well.

Some metropolitan cities may have more that one C.B.D.s and which can also act like a business center for not only a city but also for a region.

In the C.B.D.s F.A.R. should be properly controlled. In modern planning C.B.D. must have provisions for future growth. To assure this, a huge open space all around could be provided and preserved. But if this is done then there is always a tendency of the formation of slums near the C.B.D. and this is to be checked. From economical point of view, leaving a huge open space all around for future development seems illogical and the cost of providing the infrastructural facilities increases tremendously.

Conurbation

Group of towns which have become contiguous by expansion and which can be regarded as a single community for certain purposes. The modern growth of cities create a monotonous character and this has caused a reaction to their further development. This development takes place very rapidly without any control over them.

Cul-De-Sac : It is a kind of a street which stops at one end and does not give access to through crossing of a residential or other areas. The dead end of the cul-de-sac is wide so that a car can turn back and its sometimes controlled by the length of the fire brigade hose pipe from the spot of water availability. Cul-de sac generally branch out from main street and are used profusely in modern design.

Development : (as per Delhi Municipal Corporation Act; 1957) Development, with all its grammatical variations, means the carrying out of buildings, engineering, minning or other operations in, on, over and under land or making any material change in any building or land and includes redevelopment.

Green Belt : The Planners often suggest Green Belt to prevent industrial smoke, pollution and industrial noise hazards by having a belt of thick leafy trees perennial in character. There are other green belts also passing through various spots in the city which cater for leisurely walkways for the inhabitants. Such strips of green belts are prominently found in the Master Plan of Chandigarh, India.

Light Plane : For any new city, or where replanning is done to a city, sufficient space is to be left between buildings to allow light and air inside them and to take care of this, some standard is fixed for each city. This standard usually comes at an angle drawn from the top of the house outwards and downwards to the ground. This angle is usually 45° to 63.5° to the horizontal. This plane should be taken not only from the front but also from the sides. The portion of the plot included in the plane should be un-built and to be kept open to sky permanently. The same result is achieved in the planning of Garden Cities by restricting the number of houses to 12/acre.

Ribbon Development — Ribbon Development is the growth of buildings along the main thoroughfares of the city. This kind of development is usually very sudden due to want of a Master Plan and its strict adherence. What is desired to be achieved by a Master Plan is frustrated by a Ribbon Development. These developments take place so rapidly that it is impossible to control it in time and after it comes up, it becomes too late to mend it.

Urban Blight: — Urban blight implies deterioration or the existence of deficiencies in the quality of structures and their immediate environments. It also implies a range of conditions which are measurable and could be compared to some standards. Urban blight may be simple in form or complex in form.

Urban blight which is simple in form occurs:

(a) Due to deterioration of the physical character of the structures, sanitation facilities, presence of rubbish, accumulation of garbage in the yards etc.

(b) Due to adverse environmental influences like noise, odor, dust etc.

(c) Due to lacking of the community facilities like schools, play grounds, public water supply and sewerage system, adequate streets and drainage facilities.

Urban Blight which is complex in form arise when the area contain a mixture of incompatible landuses, obsolete or impractical layout of lots, blocks or streets, or if the area is unsafe and having unhealthy conditions due to use of marginal lands which are subjected to flood, marsh or tidal waves.

APPENDIX

Appendix-I

Proforma for conduction of origin and destination survey

1 2 — Station / Location :

3 4 5 6 — Date :

7 — Direction : [1-Entering city, 2-leaving city]

Sheet No :
Day :
Time
Weather :
Name of observer

Time	Type of vehicles	No. of passengers	Volume of goods	Type of goods	Origin	Destination
8 9 10 11	12 13	14 15	16 17 18 19	20 21	22 23	24 25

Instruction : Type of vehicles (Box 12, 13)

Code	Vehicle Type
01	Cycle
02	Cycle rickshaw
03	Tonga
04	Two wheelers
05	Auto-rickshaw
06	Car/Jeep
07	Taxi
08	Bus
09	Truck
10	Walk
11	Any other

Type of goods (Box 20, 21)

Code	Goods
01	Building materials
02	Food grains
03	Oils
04	Industrial raw materials
05	Finished products
06	Any others
07	Empty

Appendix-II
Proforma for traffic volume survey

Day :
Date :
Time :

Survey team :
Name of road :
Direction of flow :

Cycle	Cycle rickshaw	Tonga	Two wheelers	Auto-rickshaw	Car/Jeep/ Taxi	Bus	Truck	Any other
Total								
PCU								

Appendix-III

Proforma for landuse survey

Observation Sheet

Ward No : Survey team :

Name of street/colony : Date :

S. No.	Identification number	Landuse	No. of stories	Structural status	Maintenaince level	Year of construction

Appendix-IV

Agencies in the implementation of centrally sponsored plan schemes

(1951-1985)

Plan period	Schemes	Legislations	Implementation Agencies
1951-61	1. Social Housing Schemes (1954)		State PWD, later also Housing Board & other authorities.
	2. Slum Clearance	Slum Clearance & Improvement Acts	Slum Clearance/ Improvement Board Prescribed competent Authorities as appoint-by the State Govt.
	3. Land Acquisition & Development Schemes (1959)	Delhi Development Act, 1957	Delhi Development Authority.
1961-71	1. Preparation of Master Plans	Modern Town & Country Planning Legislation by TCPO recommending the State Govt. to amend the existing laws or introduce where none exists.	Planning Authorities to prepare Master Plans in many states came up during 1970s Urban Arts Commission.
	2. Schemes for water supply & electricity generation etc.	State Planning Legislations	Water Supply & Sewerage/Sanitation Boards. Electricity Boards.
1971-80	1. Scheme for Environmental improvement of slums (1972)	-	Being implemented by various agencies including Municipal Govts. Impro-vement Trusts, Slum Improvement Boards, Urban / City Dev. Authorites etc.

Plan period	Schemes	Legislations	Implementation Agencies
	2. I.U.D.P	Model Urban Dev. Law prepared by TCPO leading to the enactment of Development Authority Legislations by State Govts.	Urban/City Dev. Authorities.
		Urban Land (Ceiling and Regulation) Act (1976)	State Govts.
1980-85	IDSMT		Being implemented various agencies including Municipal Govts., Improvement Trusts, Slum Boards, PWD Urban Development Authorities.
1985	do		— do —
	Urban Basic Services		Program Coordination accountable to Collector Officer of Municipal Govt.; accountable to Program Coordinator.

URBAN MANAGEMENT

Institutions / Local Authorities

- (i) Urban Local Bodies
  - (a) Municipal Corporations
  - (b) Municipal Councils/Boards/Committees
  - (c) Town Area Committees
  - (d) Notified Area Committees
  - (f) Cantonment Boards
- (ii) City Improvement Trusts
- (iii) Special or Single Purpose Bodies
  - (a) Housing Boards
  - (b) Water Supply Boards
  - (c) Sewerage Boards

(d) Electricity Boards

(e) Slum Clearance and Improvement Boards.

(iv) City Planning and Development Authorities-

(v) Municipal Council

URBAN INFRASTRUCTURE SERVICES BEING PROVIDED BY CENTRAL/STATE GOVERNMENTS

1. Transport
2. Communication — postal, telephone etc.
3. Roads
4. Banking, Trade and Commerce
5. Health
6. Education
7. Social Welfare
8. Industrial Estate
9. Financial subsidies and fiscal incentives for investment in backward areas.

Appendix-V
Functions of Municipalities

KARNATAKA

Obligatory :

Construction & maintenance of Public streets, places and buildings, extinguishing fires and taking measures against fire, regulating dangerous or offensive trades or practices, removing encroachment in public places, acquiring, maintaining and regulating the places for the dead, protected water supply, naming and numbering houses, vital statistics, public vaccination, providing for animals, providing for public health and medical facilities for mental patients etc.

Discretionary :

Laying out new public streets, constructing, establishing and maintaining public parks, gardens, libraries, museums,mental hospitals, office, rest houses etc., construction and maintenance of sanitary houses, encouraging ideal arts and industries, plantation, surveys, providing entertainment, promotion of public health and child welfare, supply of necessary things during scarcity, housing, maintaining orphanages, construction and maintenance of warehouses dairy farms, transport etc.

ANDHRA PRADESH

Obligatory :

Maintenance and improvement of public streets, lighting, providing for burial, medical relief for poor, providing school education for poor, vital statistics, vaccination, preventive measures, drainage, water supply, sewage disposal, naming streets and numbering houses etc.

Discretionary :

Higher education, executing development work in the field like public streets, bridges, subways, water supply beyond the municipal limits, environmental improvement, regulating milk trade etc.

PUNJAB

Obligatory :

Maintenance of public streets, scavenging, preparation of compost, regulating offensive trades or professions, removing encroachments in public places, school education, disposal of dead animals, vital statistics, measures against dangerous diseases, etc.

Discretionary :

Executing works in various field for development of public health and sanitation, controlling pollution, drainage, regulating constructions, levy of taxes, water supply etc.

KERALA

Obligatory :

Maintenance of public streets, lighting, water supply, vital statistics, precautionary measures against dangerous diseases, regulating offensive trades and practices, sanitation, education, public health etc.

Discretionary :

Construction of water works, acquisition of lands and buildings for improvement of streets, precautionary measures against fires, dangerous trees, tanks etc., primary education, maintenance of public libraries, reading rooms, preparation of various improvement schemes and projects, etc.

MAHARASHTRA

Obligatory :

Maintaining public streets, measures against fires, regulating offensive trades and practices, removing encroachment in public places, water supply, vital statistics, naming streets and numbering houses, public vaccination, providing school education, public medical relief and organizing family planning centers, erecting municipal boundaries, welfare measures for Scheduled Casts and Scheduled Tribes etc.

Discretionary :

Establishing and maintaining public hospitals, educational institutions, libraries, museums, lunatic asylums', houses for disabled persons, town halls, rest houses etc., public parks and gardens, taking census, sanitation etc.

UTTAR PRADESH

Obligatory :

Erection of boundary marks, regulating offensive, dangerous trades or practices, removing stray dogs and obstructions in public places, acquiring, maintaining and regulating places for disposal of dead, construction, altering and maintaining public streets, culverts, markets, latrines, public drains, drainage and sewerage works, maintaining works for public health, protected water supply, public vaccination, vital statistics, primary school education, protection against fires, maintaining finances of the board, fulfilling the obligations of the Act etc.

Discretionary :

Laying out new public streets, and acquiring land for that purpose, and executing Master Plans, constructing and maintaining Public Parks, gardens, libraries, museums, leper homes, orphanages, creche and rescue homes for women, asylums, washing places, tanks, wells etc. promotion of education, conducting of census, relief works in times of local calamities, arranging for disposal of sewage, preparation of compost, constructing tramways, promoting tourism, holding fairs and exhibitions, promoting trade and industry, supply of milk etc.

RAJASTHAN

Obligatory :

Construction and maintenance of public streets, lighting public places and streets, extinguishing fires and protecting life and property when fire occur, regulating offensive and dangerous trades, removing encroachment in public places, acquiring, regulating and maintaining places for the disposal of the dead, constructing public latrines, water supply, naming streets and numbering houses, vital statistics, establishing and maintaining cattle ponds, etc.

Discretionary :

Acquisition of land for public purposes, constructing and maintaining public parks, libraries, gardens, museums and reading rooms, sanitation, planting trees, conducting of census, disposal of sewage, providing music, public health and child welfare, public entertainment, markets, holding fairs and exhibitions, supply of milk, ambulance service, establishing primary schools, housing orphans, etc.

BIHAR, ORISSA & TAMIL NADU

Obligatory :

Construction, maintenance of public roads, regulation of private roads and buildings, removal of encroachments in public places, removal and disposal of sewage, preparation of compost/manure, sanitation, regulating burials and prohibiting certain offensive trades and practices, public vaccination, water supply, vital statistics, education, etc.

Water Supply, lighting public streets, providing and maintaining drainage system and public latrines, sewage disposal, maintenance and improvements of public roads, numbering and naming of new public streets, regulating building activity, sanitation, providing for slaughter houses, regulating milk trade, vital statistics and prevention of dangerous diseases, compulsory vaccinations etc.

Discretionary :

Providing and maintaining public latrines and urinals, construction of public drains, regulation of public bathing places, extinguishing and prevention of fire, establishing municipal markets, regulating slaughterhouses, milk supply, undertaking water supply, lighting, drainage and sewerage schemes etc.

Numbering of houses, regulating the industrial activities, constructing and maintaining municipal markets, providing for public parking places and cart stands, providing for parks and gardens etc.

MADHYA PRADESH

Obligatory :

Lighting public streets and buildings, cleansing public streets, sewers and disposing of night soils and rubbish, maintenance of fire brigade for extinguishing fire, regulating and abating dangerous and offensive trades and removing projections in the public streets and places, water supply, public vaccination, vital statistics etc.

Discretionary :

Acquisition of land for public purpose, constructing, maintaining public parks or gardens, library, museum etc., taking census, constructing and maintenance of public streets, children's home, home for destitute.

DELHI

Obligatory :

Construction, maintenance and cleansing of drains, public latrines and other convenience, scavenging, removal and disposal of filth, rubbish or obnoxious or polluted matter, water supply, prevention of dangerous diseases, vaccination, vital statistics, regulation and abatement of offensive trades or practices, lighting and naming of public streets, removal of dangerous buildings.

Discretionary :

Establishment and maintenance of and aid to libraries, museums, art galleries, stadiums etc., taking census of population, construction and maintenance of rest house, poor houses, children's home, shelter for destitute and for persons of unsound mind, acquisition of movable or immovable property for public purposes.

EXCERPTS FROM SOME PAPERS PUBLISHED BY THE AUTHOR

PAPER-I

Present practice of environmental impact assessment and methodologies adopted

(Paper presented in National Seminar in 'Development and The Environment', organized by Institute of Standards Engineers, Bhopal Section, and Bharat Heavy Electricals Limited, Bhopal. India, 1994.)

1. INTRODUCTION :

Man and nature are inseparable from one another. Since the evolution of Mankind, nature has nurtured and protected man. Thus, we find that most of the ancient civilizations of the world gave due respect to mother nature by attaching some religious stigma to the various force of nature; as seen in the early civilizations of India, China, Greece, Rome etc.

It becomes evident from the old texts and mythological stories of these ancient civilizations that man knew the importance of nature and thus developed himself in the natural environments as his partner and not as an opponent for his developmental activities.

From the late 17th. Cent., with the invention of machines, man thought to conquer nature by his skill and ability to develop machines. But what he thought of the betterment of humanity, was found to be 'disastrous' in a very short period of time, not only because man was secluded from nature but also due to the fast dilapidation of unreplaceable natural resources (like fuels, minerals etc.), extermination of vast areas of forest covers, and unlimited growth of human population.

It is a well established fact that this 'lust' to conquer nature lead to the extension of many animal species thus affecting the ecological balance between the organisms. Indiscriminate felling of trees lead to a drastic change in the climate of this planet which is harboring all the plant and animal population for millions of years.

2. CLASSIFICATION OF ENVIRONMENTS :

It was during 1970s (i.e.; 1972 Stockholm Conference) that the degradation of environment due to rapid industrialization and other developmental activities started concerning the planners and as a result, various 'types' of environment were enumerated and the relationship between human beings and trie various environmental factors were studied to find out some way to carry on with the developmental activities along with the protection of the environment.

The typical classification of the environments could be stated as follows:

1. Based on Primary nature of Environmental variables; like
   (a) Exchange of energy
   (b) Exchange of matter
   (c) Exchange of information.
2. In terms of weather, food and place of living.
3. According to Adaptive Integration, i.e., how the living organisms adopt in the environment. This classification is ecological in nature and can again be sub-divided into :
   (a) Physical Environment (which within itself incorporates Biotic, Abiotic and . social Environment),
   (b) Chemical Environment.
4. Based on perceived and real world like :
   (a) Psychological Environment.
   (b) Ecological (Realistic) Environment.
   (c) Behavioral Environment.
5. Based on Natural Environment and Man-made Environment, like Social Environment, Technological Environment etc.
6. Based on (traditional) Economic Resources, like:
   (a) Renewable resources
   (b) Non-renewable resources.
7. Based on various 'spheres' around, below and above the earth; like:
   (a) Atmosphere
   (b) Bio-sphere
   (c) Hydrosphere
   (d) Lithosphere
   (e) Pedosphere etc.
8. Based on Metamorphic processes related to human beings.

[N.B. The relationships between the Human Organisms and Environmental Factors may be seen in Table-I].

9.Based on Physical Environment, like:

(a) Geographical

(b) Ecological and

(c) Behavioral Environments.

10.Based on Human Needs; like:

(a) Demographic

(b) Protective

(c) Educational

(d) Political

(e) Recreational etc.

11.Based on Area Type; like:

(a) Rural Environment

(b) Urban Environment

(c) Regional Environment etc.

12. Based on Scales; like:

(a) Macroscale

(b) Mesoscale

(c) Microscale etc.

The influence of the components of the various types of environments at different levels of human habitation mat be seen in Table-II.

TECHNIQUES OF ENVIRONMENTAL IMPACT ASSESSMENT :

Various 'Developmental Projects' has some effect on the various levels of Human Habitation and it is a felt necessity of date to have an Environmental Appraisal of those so that we can better understand the implication of a particular action. The main objectives of appraisal are:

(i) Quantification of the Implication.

(ii) To find out the 'Range' of implication.

(iii) To enumerate the Spatial dimension.

(iv) To fix up the Time dimension of the Action.

The assessment should also evaluate and communicate the relative trade-offs possible between various alternatives.

Thus we can restrict the deterioration of the environment or change it as per our requirement in a logical manner, to make this planet a more habitable place to live in. This can be done in three major ways namely:

(a) EIA (Environmental Impact Assessment)

(b) EIS (Environmental Impact Statement)

(c) EMP (Environmental Management Planning).

The most common methods of EIA and EIS with their merits and demerits are listed hereunder:

1. Expert Committee Assessment Technique.
2. Assessment by Checklists.
3. Assessment by Matrices.
4. Benefit/Cost Analysis Technique.
5. Land Unit Analysis.
6. Input-Output Overlay Techniques.
7. Cartographic Overlay Technique.
8. Information System Analysis.
9. Analysis by Mathematical Surfaces.

1. EXPERT COMMITTEE ASSESSMENT TECHNIQUE :

1.1 Concept :

To assemble a number of academic or technical experts from various fields and requiring them to write a report describing the various factors/impacts they identify.

1.2 Steps :

(i) To identify experts related to the problem.

(ii) To get reports from each expert on impacts.

(iii) To identify areas of routes of minimum impact by consensus.

1.3 Merits and Demerits :

(i) It is very simple.

(ii) It is very subjective.

(iii) It is strongly influenced by experts or any one expert.

(iv) It is not consistent from one study to another.

2. ASSESSMENT BY CHECKLISTS :

2.1 Concept :

To use some common items to be studied by experts.

2.2 Steps :

(i) To identify and list items by types/subtypes.

(ii) Marking all values.

(iii) Aggregation of marks.

(iv) Conclusion.

2.3 Merits and Demerits :

(i) It is easily replaceable.

(ii) It is easier to work with as it is in two dimension only.

(iii) The comprehensiveness of the list is always debatable.

(iv) Subjectiveness is higher as compared to some other methods

(v) Secondary or tertiary impacts are not assured.

3. ASSESSMENT BY MATRICES :

3.1 Concept :

Use of multidimensional checklists.

3.2 Steps :

(i) To make a checklist of items and sub-items.

(ii) To identify the Impact Aspects of the identified items.

(iii) To decide the method of evaluating/assessing.

(iv) To evaluate.

(v) To assimilate and simulate.

3.3 Merits and Demerits :

(i) It is more refined and widely accepted.

(ii) It is used as an input to benefit/Cost Analysis.

(iii) It acts as an aid to network techniques.

(iv) Secondary and Tertiary Impacts may be assessed by this method.

(v) The comprehensiveness of the technique us sometimes debatable.

(vi) It is not spatial.

4. BENEFIT/COST ANALYSIS :

4.1 Concept :

In this system various Impacts are identified, 'valued' and compared to expected discounted costs of project development.

4.2 Steps :

(i) To make a checklist of benefits of project.

(ii) To decide the value of the benefits.

(iii) To make a checklist of the cost of the Items.

(iv) To decide the value of the 'cost' in time.

(v) To get the discounted values of the benefit/cost.

4.3 Merits and Demerits :

(i) All the projects and their impacts being converted to monetary items, it becomes very easy for the government officials to understand.

(ii) It is very good for major projects.

(iii) A listing of comprehensive items are necessary, if the listing is not comprehensive, the benefit/cost may not be truly reflected.

(iv) Correct evaluation of the Secondary impacts and the side effects of the project are difficult.

(v) Assigning values to the benefits and costs becomes difficult due to tangible and non-tangible character of the impacts.

(vi) There is high degree of subjectivity.

5. LAND UNIT ANALYSIS :

5.1 Concept :

The approach is to assemble a team of specialists who performs a detailed inventory of significant plant communities, wild life habitats, recreational and cultural sites, established human land-use patterns and so on. They study the terrain topography and the geological nature isolating the major landscape types. An analysis is then performed to associate spatially the inventoried significant communities, land uses etc. with landscape type to derive a set of landscape units.

5.2 Steps :

(i) To decide the criteria for the landscape units.

(ii) To map the criteria and decide the units or degree of the Impacts.

(iii) To identify the potential / degree of Impact of each unit.

5.3 Merits and Demerits:

(i) Spatial dimensions are introduced in EIA.

(ii) This method may be useful for other techniques.

(iii) It may be a little difficult to select the criteria.

(iv) Sometimes the spatial data are not available and this causes difficulty in measuring the Impact.

(v) To delineate the various Landscape area units, a lot of subjectivity may be involved.

6. INPUT-OUTPUT ANALYSIS :

6.1 Concept:

The interdependence of Impact Area is worked out using the items of input and output to the project.

6.2 Steps :

(i) To make a table with inputs (with items and sub-items) along an axis; and to add outputs (with items and sub-items) along the other axis.

(ii) To add values to inputs and outputs to get the value of the Impact (in monetary terms).

6.3 Merits and Demerits :

(i) It is best used for Macro-scale projects.

(ii) Interdependence of Regions/Projects are the main focus.

(iii) Detailed data of input and output may not be got and also the data of their relationship may not be exactly known.

(iv) It is not replaceable.

7. CARTOGRAPHIC OVERLAYS :

7.1 Concept :

To use Choropleth maps for EIA.

7.2 Steps :

(i) To identify the aspects of the mapping of Impact. These aspects may be of two types: Qualitative and Quantitative,

(ii) Mapping for each aspect.

(iii) Overlaying the aspects manually or by computers.

(iv) Aggregation of these overlays.

7.3 Merits and Demerits :

(i) Space is taken into consideration in the basic form.

(ii) Even qualitative aspects may be easily considered.

(iii) Comprehensive listing of aspects is not possible.

(iv) Spatial definition of these aspects has to be generalized.

(v) The aggregation may be subjective.

8. INFORMATION SYSTEM ANALYSIS TECHNIQUE :

8.1 Concept :

The study area is divided into smaller parts; and the characters of those areas are studied by remote sensing and the analysis and book-keeping is done by computers. There are generally three types of remote sensing information:

(i) Small Format Photography (SFP).

(ii) Aerial Photography (where the details are less than that found in SFP).

(iii) Satellite Images.

8.2 Steps :

(i) To decide the type of remote sensing to be used according to the area and size of the project.

(ii) To decide the unit size.

(iii) To decide the aspects to be studied.

(iv) To decide the range of study.

(v) To aggregate the results of the studies with the help of computers.

8.3 Merits and Demerits :

(i) Subjectivity and 'human error' is less.

(ii) The process is fast.

(iii) Generally computers are mainly used as book-keeping instruments and its potential is not fully utilized, more complex assessment models for computers, utilization of logical models and threshold analysis is a felt necessity.

9. ANALYSIS BY MATHEMATICAL SURFACES :

9.1 Concept:

To use some derived/ proved correlation in mathematical form for regions.

9.2 Steps :

(i) To identify the key and secondary variables of the Impact.

(ii) To evolve a mathematical 'model' of the variables through available literature or by empirical derivative work,

(iii) Using the mathematical models various Impact values are worked out for various sub-areas of the study area.

9.3 Merits and Demerits :

(i) The process is highly quantified and very convincing

(ii) Everything cannot be quantified exactly.

(iii) All thing cannot be correlated.

(iv) Some level of subjectivity is still there.

(v) The weights are highly based on assumptions.

ENVIRONMENTAL ASSESSMENT : RELATED PROBLEMS

In the preceding review, problems related to specific environmental assessment methodologies were outlined. In addition to these problems, a number of general problems may be identified.

THE VALUE RANKING PROBLEM :

The fundamental problem is the lack of agreement on appropriate methodology to use and the effects to consider. In addition, for nearly every methodology, some choice must be made concerning the parameters to analyze. Indirect effects are rarely included in the process. Moreover, once the effects have been selected, the measurement and scaling of the effects require further judgment. Often, monetary values are attached to intangible items, inviting criticisms.

THE NO DEVELOPMENT PROBLEM :

No process (apart from the Benefit/ Cost techniques) appears to include a specific consideration of whether or not to go ahead with the project. When routing a utility, it is obvious that certain sub-areas of the study area must be considered as 'No-Go' or 'No-development'.

To be effective, assessment and routing methodology for utilities at a sub-regional level must permit a project level assessment of the No-Development problem. If no routes can be located which could avoid the No-Development area, then, the No-Development status of the total project may be determined.

STUDY COMPREHENSIBILITY AND PUBLIC INVOLVEMENT :

Most environmental assessment are destined for public review under governmental requirements. This requires the result of the analysis to be easily comprehended and the inclusion of public concerns and perceptions in the study. Most assessment techniques do not provide for accommodation of public concerns and perceptions. Input-Output Analysis is likely the worst offender with its massive tableau of technical coefficients. Benefit/ Cost Analysis is not much better with its use of discounted weighted monetary value assessment of intangible parameters. The cartographic overlay technique appears to provide the most comprehensive results especially if supporting analysis maps are provided. The actual direct use of data reflecting public concerns is rarely encountered.

ANALYSIS SCALE AND PLANNING REQUIREMENTS :

The nature of each environmental assessment methodology establishes a 'scale' or 'level' at which its application is more appropriate. Many studies involving checklists, matrix methods, benefit/cost analysis and input-output analysis are high level assessments (Macro Level analysis) across the study area. Mathematical surface approximation techniques make use of trend surface fittings and smooth data. In other words, the technique generalizes and is most appropriate for higher level studies. Only the cartographic overlay and computer based information systems appear to be fully capable of low level evaluation.

It is essential that a detailed localized study be performed if a realistic E.I.A. is to be made. The implication is that a proper high level of policy level assessment of environmental impact cannot be made without an adequate low level assessment first being performed to provide an accurate environmental cost estimate.

While the cartographic methods and computer based information system show promise for a detailed low-level analysis, the scale of the important physical phenomena gives rise to difficulty.

STUDY FLEXIBILITY :

The ongoing development of the environmental standards and increased public involvement in assessments gives rise to adjustments being required to study methodology either in the analysis or in the review phase. This requires an analysis method which can relatively easily accommodate added (or deleted) data as well as changes in the valuation of weighing of the analysis model. Few of the established methods contain much of this flexibility. To be fully successful, a methodology must not only provide assessment of alternatives but also be capable (during the study phase) of assessing alternatives proposed by interveners during the review stage. This calls for added flexibility in the methodology.

GENERAL LIMITATIONS OF ROUTING METHODOLOGIES :

In the previous discussions limitations of specific path-finding methods have been outlined. But there are some more general limitations. Routing a utility across even a moderate study area and maintaining accuracy are difficult to achieve. A method is needed to reduce the complexity of route finding and simultaneously maintain adequate accuracy.

CONCLUSIONS :

Thus it becomes evident that Environmental Impact Assessment is a complicated process and a lot of research work is still needed to arrive at a comprehensive methodology which should be flawless on one hand and comprehendible by the public on the other.

Inefficient administrative and managing machinery of the governmental organizations along with the vested political interests also make environmental impact assessments useless in most of the cases. On the other hand, if non-governmental organizations are given the responsibilities of assessing the impacts on the environment care should be taken that they do it with a genuine intention to protect the living organisms and the environment, as the chances of personal influences are likely to be more in these cases. The NGOs should also have a proper team to assess the environment and must be equipped with all the modern machineries (computers etc.) and technicalities.

To create a comprehensive environment, assessment of environmental impact is necessary at all levels for which time, money and data sources are three main important things to be taken care of. If any of these are not supplied in a proper manner, no assessment will have any meaning whatsoever.

Thus we can conclude that to protect the world E.I.A. is a must. It should be done at all the 'levels' of project.

It should have a methodology which is easy, cheap and includes all the tangible and intangible aspects of the environment.

The organizations assessing the impacts (whether governmental or non-governmental) should have a 'will' to assess along with all the resources to do it.

Unless and until we do these things, no legislation or 'act' can help to save the environment and whatever policy we adopt, they should be internationally followed as localized (or even nation-wide) policies will not help to protect the environment as environment is the property of the whole world and not of any particular nation or country. The essence of this could also be derived from the ancient Sanskrit proverb of 'Vasudhaiva Kutumbakam' or 'The Whole World is a Family' and every body must do his/ her best to save it.

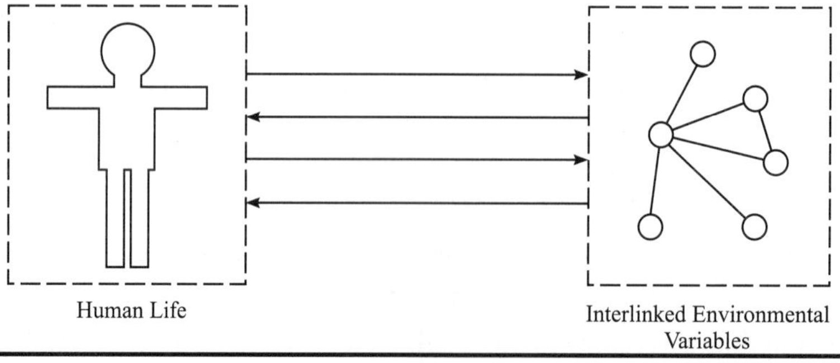

Fig. No. I : ***Man in envronmental system***

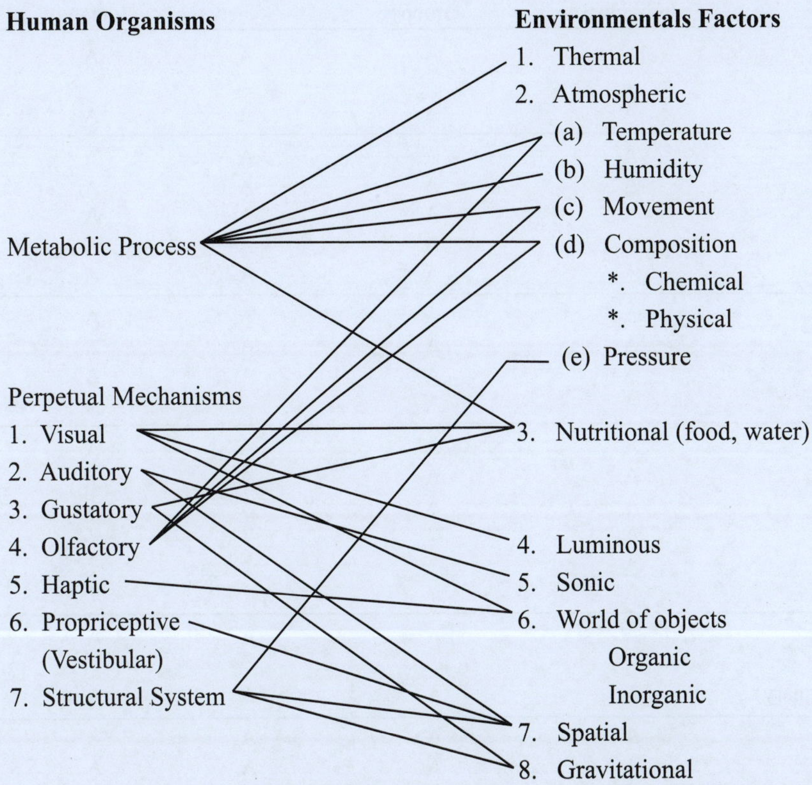

Fig. No. II : ***Relationship between organism and environmental factors***

The influence of the components of the various types of environments at different levels of human habitations

Components	Individual	Group	Society	Humanity
Urban & Rural	A	A	NA	NA
Energy	A	A	A	A
Matter	A	A	A	A
Information	A	A	A	A
Weather	A	A	A	A
Food	A	A	A	A
Place to live	A	A	A	A
Other Organisms	A	A	A	A
Physical/Chemical	A	A	A	A
Biotic	A	A	A	A
Social	A	A	A	NA
Perceived	A	?	?	?
Real	A	A	A	A
Natural	A	A	A	A
Manmade	A	A	A	A
Social	A	A	A	A

Components	Individual	Group	Society	Humanity
Resources	NA	A	A	A
Other Natural Elements	A	A	A	A
Atmosphere	A	A	A	A
Biosphere	A	A	A	A
Hydrosphere	A	A	A	A
Pedosphere	A	A	A	A
Lithosphere	A	A	A	A
Subjective	A	A	A	A
Objective	A	A	A	A
Operational	A	A	A	A
Ecosphere	A	A	A	A
Physical	A	A	A	A
Geographical	A	A	A	7
Ecological	A	A	A	A
Behavioral	A	?	?	?
Conscious	A	?	?	?
Existential	A	?	?	7
Natural	A	A	A	A
Artificial (May be imaginary)	A	A	A	A
Natural	A	A	A	A
Manmade	A	A	A	A
Institutional	A	A	A	A
Thermal	A	A	A	A
Atmospheric	A	A	A	A
Nutritional	A	A	A	A
Luminous	A	A	A	A
Sonic	A	A	A	A
Spatial	A	A	A	A
Gravitational	A	A	A	A
Population	A	?	?	?
Protective	A	?	?	?
Educational	A	?	?	?
Political	A	?	?	?
Recreational	A	?	?	?
Ideological	A	?	?	?
Values	A	?	?	?
Natural Resources	?	?	?	A
Habitational Conditions	A	A	A	A
Economic & External	A	A	A	?

N.B. The component groups are made as per objectives of study.

INDEX:

A = Applicable
NA = Not applicable
? = Debatable.

PAPER-II

Unintended metropolitan growth in India : A methodology for its environmental management

[Published in Journal of Pt. Ravishankar Shukla University, Raipur, (M.P.) India, Vol.9; No.B (Science); 1996, pp 11-23]

1. DEFINITION OF UNINTENDED GROWTH :

Unintended growth is an unplanned and yet somewhat chaotically developed, neglected section or part of the city, over-crowded with ill-attended houses and other structures with an assortment of activities some of which may be harmful with the prevalent inadequate public utilities and services. Thus unintended growth may be described as that physical growth out of the city (or within the city limits,as the case may be,) which has not yet been perceived, provided, planned and permitted by city plan. It may be residential, commercial, industrial, or an agglomeration of all these activities.

This problem is more prominent in the expanding cities of the developing countries. The spectacular growth of Urban population in these countries is characterized by accelerating growth rate of Urban immigration propelled by distress in the countryside and pulled by the attraction of the cities.

The rural people make their own settlement on illegally occupied land as there is an inaccessibility of low cost housing and the initial allocation of land and of resources are not available. There is no other way in which large number of low income families can secure an ecological niche in the city.

In the developing countries there exists a very sharp distinction between the 'Urban Rich' and the 'Urban Poor'. The migrated urban poor offer a cheap labor to the Rich and gradually they are incorporated within the social framework. There exist some sort of 'symbiotic' relationship between the poor and the rich. Hence, we find multistoried apartment blocks are flanked by slums and squatter settlements, or the C.B .D. of the metro is embanked by the migrated people. A bird's eye view on any city of the developing world would reveal the truth.

8. MODES OF DEVELOPMENT OF THE UNINTENDED GROWTH : MODUS OPERANDI OF THE SPECULATORS —THE DELHI EXPERIENCE :

In the developing countries, the development of unplanned, unintended growth follows a typical mode. This could be understood by the experience of Delhi, where these lands were developed by the speculators.

Under the Land Reform Act of 1954, all the privately held agricultural land around Delhi was consolidated. The allotment of land after consolidation was given to 'Bhumidars'— who were the legal owners of the land.

The Bhumidars gave this land on 10 years' lease to cultivators (Assamee) to cultivate. Thus the land title got duplicated. In some cases the Assamee carved out plots from the land and disposed it off. There were also instances when the lease could not be renewed after 10 years thus giving the lessee a chance to become the owner.

Now, the colonizer makes a deal with the owner and he carves out the colony on the land , the owner registers the sale with the registrar and the colonizer builds a wall with one room so as to give it a built-up look. The people got the land at cheap price and are encouraged to build. The house tax is often paid from a back date.

If the local authority sends a notice to the builder, a stay-order is brought and the period of the stay order is 4 to 5 years in which time 50% to 60% of the area gets built. Politicians identified with the area bring pressures on the local authorities for gradual improvement of the area which are gradually very slow and less beneficial.

Before the stay is vacated, the colony is already functioning, and these areas are identified as 'Jhuggi-jhompri' or 'Busti' areas. During the preparation of the Master Plan of Delhi, the D.D.A. identified the major 'Busti' areas of Delhi and calculated the distribution of population in them in each of the planning divisions. The density of the 'Bustis' were greatest in A and B divisions ranging from 421 to 1634 ppa (persons per acre);while in the C and D divisions it is from 129 to 613 ppa and thins down to 123 to 27 ppa in the E and F divisions. The congestions of these 'Bustis were further revealed by the fact that about 60% of the dwelling units are built on 45% of the total area under Bustis and these units average more than a 1000 per Busti.(Majumdar.T.K.; 1980)

A brief description of the situation of the squatter settlements of Delhi will make the point clear :

According to Muddassir,(1993) as per the latest estimates there are about 900 Jhuggi-Jhompri clusters in various areas of Delhi having a population of around 15 lakhs.

A survey of 352 J.J. Clusters in May 1989 showed that there were 138 J.J. clusters which had no source of safe water within the cluster. There were 1,032 taps available which means 220 households per water tap. 536 households were sharing one deep hand pump. Taking these two sources together, there were 156 households per safe water point.

The provision of safe water sources is highly inadequate even if judged by the norms laid down by the Environmental Improvement Scheme (one tap per 100 families).

The same study shows that there were one latrine per 31 household, when the EIS recommendations are 30 household per seat.But the gravity of situation can be comprehended, if we see the spatial distribution of latrines. It was found that there were no community latrine or Sulabh Sauchalaya in 73% of the J.J. clusters in south Delhi, 65% in the west, 62% in the north and 40% in east Delhi.

The Comptroller and Auditor General of India reported that out of 251 J.J. clusters in east Delhi surveyed in December 1988, 250 had no drainage facility.

These illustrations proves that level of services provided and the quality of life of these settlements are very poor and thus the environmental condition is also not congenial for living.

3. THE CONSEQUENCES :

The colonies are developed just to house the migrants but no planning or thought was given towards the development of the environment of these people and no study is also done to evaluate the impacts those settlements are having on the environment. There are many factors which can be identified having some direct and/ or indirect effects on the environment at the macro and micro levels, on which the Q.O.L.(Quality of Life) of the people depends.

The Q.O.L. is the quality of standards (or indicators) of the inter-connected entities of the given environment. Q.O.L. depends upon indefinite conditions and the 'indicators' should be selected to achieve the 'goal' and to fulfill the 'purpose' for which they are chosen. There could not be any universal set of criteria for evaluating the Q.O.L., as many of them could be subjective (related to education, housing, experience, knowledge etc.);but the evaluation has to be objective in nature. These may be:-

1. Economical (i.e.accumulated assets,cost of living, consumption pattern etc.)
2. Social.psychological, cultural.(i.e. political process, public participation, education, social relations, health etc.)
3. Environmental, physical (i.e. pollutions, depletion of natural resources, housing and shelter conditions etc.).

The factors affecting the Q.O.L. may broadly be classified into:

(a) Physical: Lack of water supply, electricity, sewerage, garbage collection, light, ventilation, standard livable sizes of rooms, overall aesthetics etc.

(b) Social: Poor Q.O.L. lead this settlements to become the 'breeding grounds' of social vices, like drug addicts, smugglers, anti-socials etc., sometimes with the backing of the politicians with vested interests, which spread their net slowly and gradually over the Urban fabric, which does not need any statistical data to make it more clear to the people of the world today.

Though these are long term consequences, planners and sociologists feel that the improvement of the environment alone can create a better Q.O.L. and consequently better citizens.

People, in general, does not give environment the first preference in their personal priorities and hence, this leads them to a worse condition of life which makes the society as a whole degrading and economically less productive. The placement of Environment in personal priorities may be empirically shown as given in Table-1. It seems education and environmental consciousness is the need of the hour amongst the Urban population, both rich and poor.

4. PROBLEM IDENTIFICATION AND EMERGING POLICIES :

The problems of environment in unintended urban growth may be identified under the following heads :

1. Relationship amongst the various landuses, housing conditions, provisions of other physical and social infrastructures (like health, education, recreation etc.).
2. Traffic generations (w.r.t. quantity, quality and intensity).
3. The inputs (i.e. the factors that help in the growth of the developments).
4. The outputs (i.e.wastes, garbage, sewerages etc.).
5. Soil conditions/ground relation.
6. Land values and other economical aspects.
7. Economic developments and safety.
8. Microclimate of the settlements.
9. Aesthetics of the overall area.

After analysing the problems under these aspects, the most appropriate method of E.I.A. may be chosen and then applied in a proper manner and managed in the said methodology (vide Table-II) to give the overall picture of the environment.The method chosen for E.I.A. must be cheap, comprehendible and the implementation machinery must be sound and efficient enough to give the results of the implementation within a very short time.Thus any method to improve the environment must aim at the economical aspects of the people and they must be educated to improve their own environment for a better condition of life.

Then only a proper 'data base' may be made for the 'feed back' to check the results of the implementation.

The major hurdles in the Environmental management of the Unintended growths may come as:

(i) Use and selection of planning standards.

(ii) Political and legal interventions.

These are to be taken proper care of during the formation of the implementation methodology.

5. PROPOSED METHODOLOGY FOR ENVIRONMENTAL MANAGEMENT PLANNING (E.M.P) :

After assessing the magnitude of the deterioration of the environment of any such unintended growth, a methodology involving three logical steps are to be followed. These steps are:

(i) E.I.S.: Environmental Impact Statement.

(ii) E.I.A.: Environmental Impact Assessment.

(iii) E.M.P.: Environmental Management Planning.

An appropriate method of E.I.A. for E.M.P. may be applied. The common methods of E.I.A. for application of E.M.P. are listed below:-(Bandyopadhyay,A.;1994)

1. Expert Committee Assessment Technique.
2. Assessment by Checklists.
3. Assessment by Matrices.
4. Benefit/Cost Analysis Technique.
5. Land Unit Analysis.
6. Input-Output Overlay Technique.
7. Cartographic Overlay Technique.
8. Information System Analysis.(Computer Based Information System).
9. Analysis by Mathematical surfaces.

All these methods have their own benefits and difficulties and describing each of them is out of the scope of this paper. But the general problems for E.I.A. may be identified which are common to most of the techniques. They are :

(a) THE VALUE RANKING PROBLEM :

The fundamental problem is the lack of agreement on appropriate methodology to use and effects to consider. In addition, for every methodology, some choice must be made concerning the parameters to analyse. Indirect effects are rarely included in the process. Moreover, once the effects have been selected, the measurement and scaling of the effects require further judgment. Often monetary values are attached to intangible items, inviting criticisms.

(b) THE NO DEVELOPMENT PROBLEM :

No process (apart from the Benefit/Cost Technique) appears to include a specific consideration of whether or not to go ahead with the project. When routing a utility, it is obvious that certain sub-areas of the study area must be considered 'No-Go' or 'No Development'.

This assessment of the 'No-Go' routing methodology at a sub-region level would permit the project level assessment of the 'No Development' problem. If no routes can be located which avoid 'No Development' area, the 'No Development', status of the total project may be determined.

(c) STUDY COMPREHENDIBILITY AND PUBLIC INVOLVEMENT :

Most environmental assessment are destined for public review under governmental requirements. This requires that the results of the analysis to be easily comprehended and the inclusion of public concerns and perceptions in study. Most of the assessment techniques do not provide for accommodation of public concerns and perceptions. Input-Output analysis is likely the worst offender with its massive tableau of technical coefficients. Benefit/Cost analysis is not much better with its use of discounted weighted monetary value assessments of intangible parameters. The Cartographic Overlay Technique appears to provide the most comprehensive results especially when supporting analysis maps are provided. The actual direct use of data reflecting public concerns are rarely encountered.

(d) ANALYSIS SCALI AND PLANNING REQUIREMENTS :

The nature of each environmental assessment methodology establishes a scale or level at which its application is most appropriate. Many studies involving checklists, matrix method, benefit/cost analysis and input-output analysis are high level analysis (Macro Analysis) across the study area. Mathematical surface approximation technique make use of trend surface fitting and smooth data. In other words, the technique generalises and is most appropriate for higher scale studies. Only cartographic overlay and computer based information systems appears to be fully capable of low level evaluation.

It is essential that a detailed localised study be performed if a realistic E.I.A. is to be made. The implication is that a proper high level assessment of environmental impact cannot be made without an adequate low level assessment first being performed to provide an accurate environmental cost estimate.

While the Cartographic methods and Computer Based Information System shows promise for a detailed low level analysis, the scale of the important physical phenomena gives rise to difficulty.

(c) STUDY FLEXIBILITY :

The ongoing development of the environmental standards and increased public involvement in assessments gives rise to adjustments being required to study the methodology either in the analysis or in the review phase. This requires an analysis method which can relatively easily accommodate added (or deleted) data as well as changes in the valuation of weighing of the analysis model. Few of the established methods contain much of this flexibility. To be fully successful, a methodology must not only provide assessment of alternatives, but also be capable to incorporate and assess the alternatives proposed by interveners during the study and the review stage. This calls for added flexibility in the methodology.

(f) GENERAL LIMITATIONS OF ROUTING METHODOLOGIES :

Routing a utility across even a moderate study area and maintaining accuracy is difficult to achieve. A method is needed to reduce the complexity of route finding and maintain adequate accuracy.

However, the pros and cons of each method is required to be studied in detail before application and may be made applicable for E.M.P. for any such area. Table No.-III includes a complete methodology for the E.M.P. of the unintended growth area.

The E.M.P. should also incorporate the following characteristics:

(i) It should define quality standards.

(ii) It should do periodic assessment of standards and achievements.

(iii) It should do monitoring and give timely warning to the violators of the proposed regulations.

(iv) It should involve public participation to take care of the delays that takes place due to political and legal interventions.

(v) It should be empowered to prepare 'Action Plans' as preservative measures.

(vi) It should adopt Urban Renewal/Conservation techniques as curative measures.

(vii) It should be empowered to implement fines, taxes etc. as punitive actions.

Public Participation to resist political interventions with vested interests have been incorporated in the proposed methodology to make it more functional and implementable.

6. CONCLUSION :

Hence, to make the environment of the unintended growth of metropolitan cities, a holistic approach should be taken for the development of the overall Q.O.L. of the people.

The problems of these unintended growths are multifaceted and multi-dimensional in form and dynamic in nature. The problems are also linked with one another and they change in time, which makes the identification of these more difficult; for which an efficient

E.M.P. is required to be implemented.

The E.M.P. should be a concious systematic effort to produce aesthetically pleasing, economically viable, and physically healthy environment. This should cover:-

(i) Natural and man-made aspects like air, water, noise, sewerage, solid waste etc.

(ii) Energy conservation.

(iii) Preservation of sites of historical importance.

(iv) Creating aesthetics.

(v) Health, housing, mass transportation and recreation.

(vi) Economical development, transportation, safety etc.

If the proposed methodology incorporates these factors and if any serious thought is given for its implementation it will emerge out as an efficient E.M.P. Methodology. But it should be first tested on a small scale and then adopted for larger projects.

REFERENCES:

1. Majumdar, T.K.;(1980) " Urbanising the Poor, a Sociological Study of Low Income Migrant Communities in the Metropolitan City of Delhi":Lancers Publishers, New Delhi.
2. Muddassir,S.M.(et al);(1993) "Health-Infrastructure Relationships in Squatter Settlements of Delhi", Delhi, Ed.Institute of Town Planners, India, [ITPI Journal vol- 12, No.2(156)].
3. Bandyopadhyay, Abir;(1994) "Present Practice of Environmental Impact Assessment and Methodologies Adopted". Report of National Seminar on Development and the Environment, BHEL, Bhopal.
4. Devuyst, D.; 1992, Internal Report, Department of Human Ecology, Vrije Universiteit, Brussels, (Belgium). PA

Table I

Placement of environment in personel priorities

1. Economic
   - Job → Income → Work satisfaction
2. Social
   - House → Education for children → Income differentials → Consumption of materials & amenities
3. Friendly neighbourhood → Community participation → Universal justice

Note : Step 2 involves environment directly
Step 3 involves environment indirectly

Table II

Proposed E.I.S. Procedure

Project Proposal

↓

Screening

↓

Preparatio of Draft E.I.S.

↓

Review and Public Participation

↓

Preparation of Final E.I.S.

↓

Final Review

↓

Decision

↓

Evaluation and Monitoring

Source : Devuyst, D.1992

Table III

Methodology for Environmental Management Planning for Unintended Growth

1. Define problem	(a) Study of the history of development and growth pattern.
	(b) Survey for information regarding the Q.O.L.—and identification of appropriate environmental indicators.
	(c) Define objectives of environmental improvements by public participation and their priorities.
2. Identify and analyse the possible actions	(a) Review of possible actions by establishing suitable standards of the indicators as identified in step l.b.*
	(b) Doing the cost/benefit analysis.
3. Prepare draft plan	(a) Long term Plan (to uplift the socio-economic standards),
	(b) Short term Plan (Action Plans).

4. Public participation & opinion	(a) Draft Plan made available to public.DoE, NGOs etc.
	(b) Arranging for community workshops, public meetings and getting Public Opinion.
	(c) Collecting supplementary Public Opinions by Primary/Secondary data.
	(d) Getting opinion of experts
5. Finalise plan	(a) Long Term Plan
	(b) Short Term Plan (Renewal Programmes, Action Plans etc.)
6. Implement plan	(a) Initiate Planning process.
	(b) Involve" Public Participation.
	(c) Administer Development Process.
	(d) Check Atrocities by punitive measures.
7. Evaluate responses and adjust actions by periodic assessment	(a) Survey for responses.
	(b) Create Data Base.
	(c) Modify quality of standards.
	(d) Modify Plans.

*.The possible may be': (a) Renewal, (b) Replacement, (c) Relocation, (d) Modification, (e) Supplementation, (f) Enclosure of Sources, (g) Erect Intersectors, (h) Shedding the Receiver, (i) Personal Protection.

PAPER III

Town planning a socialistic approach—The Chinese experience

[Published in the Journal of Indian Institute of Architects, Vol.61; Issue 05; August 1996; pp. 15-17].

INTRODUCTION :

Town Planning is always said to be on the cross roads. For the various problems that come on its path, it had always been a practice of experimenting with any of the selected theoretical solutions and the experiencing its effects in due course of time. Often the 'feedbacks' got are so bitter that planners are blamed for not choosing the right solution for the problems at the right time. People realize the effects of a bad planning much later and when the whole problem becomes too cumbersome to tackle with any further. The reason for this is simple—the theories that are applied depends on many factors like political, social, economical etc. and hence a direct bearing on the people of a particular country and a particular place. Thus the borrowed theory of one country are hardly of any help to any other country or society. In other words, no universal blueprint is possible to be made and applied to ameliorate the problems of an Urban area.

In India, we have practiced the British and American policies to solve our urban problems for over 40 years now, and very less we have achieved in solving those. We have failed to stop the in-migration to our metropolitan cities and to bring up the Quality of Life (Q.O.L.) of the people in both the urban and the rural areas.

Now, we are to take some very important decisions, that is, should we continue with the same policies, or should we look for something else ? again if we think of 'something else', should we search for any other precedence? If we do, then the only other option which we have is perhaps to look at China, not only because of its gigantic amount of population (like India) and is situated in Asia, but also because it adopted a totally different trend to solve its urban problems.

THE CHINESE CITIES :

China, when it became the 'People's Republic' in 1949,went under a total mutation in the economical, social and political conditions. By 1953, there were 166 big cities with 9 cities having more than 10 lakhs (10,00,000) people, 16 having population between 5 to 10 lakhs, 77 having population 1 to 5 lakhs and 64 having a population of less than 1 lakh. Most of these towns have a typical spatial structure (except Harbin), as described hereunder:

The City Structure :

The cities consisted of only 3 zones :

(i) A Central Industrial Core of continuously built-up area accommodating the elite, educated mass of people and also the industrial workers.

(ii) This was surrounded by some smaller settlements and the Seats of Communes which, along with the Central Core were under a municipal or district boundary. These smaller settlements were acting as market centers and producing centers of vegetable and non-stable food, these smaller towns were known as rural 'XEN' (County, Districts or Tehsils).

(iii) Surrounding these were the agricultural fields with a few scattered settlements of agricultural labors. Therefore the concept of 'Urban Area' in China was ambiguous as various population sizes for the same city is found. China had 299 cities having population beyond 500,000.

Hence, 2 basic models of Urbanization were followed in 1949. The socialistic ideologies of Marx-Engles was of accepting the cities as places of 'exploitation of the laborers'. Stalin pointed out that cities are the production centers of industrial and non-agricultural productions, but Marx- Engles totally denied any role of the cities.

China followed Stalin's model by developing the concentration of the Core Cities and increasing the industries. By 1965, Mao-Tse-Dong evolved his own theory to solve the deteriorating conditions of the urban centers. The main features of Mao's theory were:

(i) There should not be any difference in town and country, i.e., towns should not only be benefited by receiving the 'lion's share' of the resources.

(ii) The Central City Core is the only suitable place industrialization which requires more 'mental labor' than the agricultural fields, which, on the other hand, requires more 'manual labor'. So, the areas of more 'manual laborers (i.e. the agricultural fields) are devoid of the 'mental laborers' who concentrated only on the large cities. Hence, some 'mental laborers' are needed to be shifted from the towns to the villages. As the industrial labors are immovable, the educated people were forced to move to the villages by a special policy dictum. This Town- Country movement was known as 'SHANG SHAN XIA XIAN (HSIA HSIANG). This movement was followed in 1970-1978. The Gang of Four then took the power from the ailing Mao and then at least 12 million educated people were literally 'forced out' from the Urban Areas. The comparison of the figures in Table-I of 1979 and 1981 will prove the result.

Immediate Consequences :

The cities grew only due to natural growth (as migration was stopped forcefully) and the growth of the cities were immediately arrested. This resulted into a model of Urbanization not based on any historical pattern of urban growth and the concentration of the cities were reduced to a great extent. Moreover some British and American authors say that this practice was 'anti-urban' and 'de-urbanization'.

Far Reaching Consequences :

The Town Country movement lead to pragmatic and hard core rural development which ultimately resulted in over concentration-of the rural settlements. Thus the rural areas are slowly converted to urban areas as the educated 'mental laborers' were of no good to the agricultural 'manual laborers'. Moreover, the 'active' rural land became 'passive' when transferred to urban land. These urban areas lead to the formation of other counter points of production and because of the vacuum created at the central core of the old urban areas (due to out-migration), the industrial areas got more concentrated there, and ultimately got scattered in space near the newly formed urban areas, and these again, started developing as new 'Cores'.

Now, from 1978, in the post Mao period, China decided on liberalizing socialism by 'Open Door Policy' by which the western education was allowed to come in to give the advantage of the western technical expertise to the Chinese people. Secondly, multinational companies were allowed to business in Chinese soil to bring up the economic level of the people.

These two forces again went to the development of the new as well as the old urban centers and thus large cities grew up more to form urban agglomeration.

Hence, by 1981 China had 37 cities of more than 10 lakh people, 41 cities of population 5-10 lakhs, 136 cities had 1-5 lakh people and few only were left with less than 1 lakh population.

Feed Back :

What we have learnt from the experimentation of Mao's model of urbanization is:

1. Unless the GNP and the PCI (Per Capita Income) index of the people are not improved, any such policies are not useful.
2. Selective (and forceful) decentralization is not helpful for the overall development of the urban centers.
3. The 'Goals' of socialistic model had to be changed when capitalism (in the form of education and business) came into the picture of this new model.

Conclusion:

Nevertheless, the Chinese model was a good experiment to do and a better feed-back to us, who still could not find any amicable solution of the ever increasing problems of the urbanization in our country.

Moreover, we are also moving towards an open market economy with not so much improvements in our GNP and PCI indices.

Now, with the virtual collapse of socialism in most of the countries, and with the birth of new countries in the world map, there is still a scope to learn from the policies that are suitable for these newly formed countries.

But whatever might be the policies, will those be suitable for us? Or shall we wait to see what ultimately happens at the last social resort, e.g. China, and get another feed back? There are many options open to us, many ways promising us to our goals safe and sound. It is now upon us, the makers of this great nation, to select which path we choose, and how fast we should move towards the internationalization of our economy, which will ultimately reflect in our overall scene of urbanization.

Table I

Distribution of settlements in China

Population	Number of cities					
Year	1936	1953	1957	1972	1979	1981
1000,000+	06	09	14	21	29	37
500,000-1000,000	05	16	20	22	36	41
300,000-500,000	07	10	_	38	41	50
100,000-300,000	57	67	90	33	85	86
Less than 100,000	116	64	52	47	-	15
Total	191	166	176	181	191	229

Total small Towns= 3171 Total number of all settlements= 3400

Table II

Chinese cities in order of population : (Source : Sit; F.S.,'Chinese Cities'; 1985)

Shangahi	6,134,000
Beiging	5,430,000
Tianjin	5,020,000
Shenyang	3,918,000
Wuhan	3,157,000
Guangznou	3,077,000

Calcutta was the most populated city of the World in 1981, having a population of 9170,000. Delhi had a population of 5,710,000. It was estimated that the population of Shanghai will supersede Calcutta by 2000 A.D.; the Calcutta will rank 7th. or 8th. in the world.

INDEX

UNIT I

UNIT II

UNIT III

UNIT IV

RECOMMENDED READINGS

1. Keeble; Lewis (1976); "Principles & Practice of Town and Country Planning": The Estate Gazette Limited, London (Copyright 1969).
2. Delhi Development Authority; " Master Plan for Delhi": 1962.
3. Chattopadhya: B.C.: " History of Human Settlements: (Associateship Examination)": Institute of Town Planners, India; New Delhi.
4. Gallion; A.B. and Eisner; S.; (1969), " The Urban Pattern — City Planning and Design": Affiliated East-west Press Pvt. Ltd.; New Delhi (Copyright 1963).
5. Conyers: D. and Hills: P.: (1984). "An Introduction to Development Planning in the Third World": John Wiley and Sons, Bath; pp. 3-9.
6. Keeble; Lewis (1983): "Town Planning Made Plain": Construction Press, London;pp.1-9.
7. Chatterjee; M.N.;(1987); "Introduction to Planning — Synopsis of Lecture": School of Planning and Architecture, New Delhi (Un-published).
8. Indian Road Congress; (1983); "Geometric Design Standards for Urban Roads in Plains.— QRC: 86- 1983V: New Delhi.
9. Shukla; D.N.; (1993); "Vastu Shastra. Volume-I. Hindu Science of Architecture": Munsiram Manoharlal Publishers Pvt. Ltd.; New Delhi.
10. Bacon; Edmund N.; (1967); "Design of Cities": Thames and Hudson, London.